Helping Young Refugees and Immigrants Succeed

Helping Young Refugees and Immigrants Succeed

Public Policy, Aid, and Education

Edited by

Gerhard Sonnert and Gerald Holton

HELPING YOUNG REFUGEES AND IMMIGRANTS SUCCEED
Copyright © Gerhard Sonnert and Gerald Holton, 2010.
Softcover reprint of the hardcover 1st edition 2010 978-0-230-62196-1
All rights reserved.

First published in 2010 by PALGRAVE MACMILLAN® in the
United States - a division of St. Martin's Press LLC, 175 Fifth Avenue,
New York, NY 10010.

Where this book is distributed in the UK, Europe and the rest of
the World, this is by Palgrave Macmillan, a division of Macmillan
Publishers Limited, registered in England, company number 785998,
of Houndmills, Basingstoke, Hampshire RG21 6XS.

Palgrave Macmillan is the global academic imprint of the above
companies and has companies and representatives throughout the world.

Palgrave® and Macmillan® are registered trademarks in the United
States, the United Kingdom, Europe and other countries.

ISBN 978-1-349-38373-3   ISBN 978-0-230-11296-4 (eBook)
DOI 10.1057/9780230112964

Library of Congress Cataloging-in-Publication Data

Helping young refugees and immigrants succeed : public policy, aid, and
    education / edited by Gerhard Sonnert and Gerald Holton.
      p. cm.
    Includes bibliographical references.

    1. United States—Emigration and immigration.   2. Refugees—United
    States—Social conditions.   3. Immigrants—United States—Social
    conditions.   4. Children of immigrants—Education—Social
    Conditions—United States.   5. Social work with immigrants—
    United States.   I. Sonnert, Gerhard, 1957–   II. Holton, Gerald James.
    JV6465.H4 2010
    362.87′5610973—dc22   2010004572

Design by Integra Software Services

First edition: September 2010

10 9 8 7 6 5 4 3 2 1

To those who have been helping young refugees and immigrants succeed.

Contents

List of Tables   xi

Introduction   1
Gerhard Sonnert and Gerald Holton

Part I   Setting the Stage: The Historical Framework

1. Questions of Success: Lessons from the Last Great Immigration   9
Nancy Foner

2. Conceptual Frameworks for Obstacles to Immigrant Advancement: Perceiving Immigrants in the Context of Societal Change   23
Reed Ueda

Part II   Looking into the Past: Some Lessons from the "Second Wave"

3. Successful Young Refugees from Central Europe—Potential Lessons for Today   37
Gerhard Sonnert and Gerald Holton

4. The Second Wave and the American-Jewish Community   51
Nathan Glazer

5. Helping Young Immigrants/Refugees Become Entrepreneurs   61
Clark Claus Abt

Part III   Recent and Current Waves of Young Immigrants and Refugees

6. After the Cold War: Comparing Soviet Jewish and Vietnamese Youth in the 1980s to Today's Young Refugees   75
Steven J. Gold

7   Second Generation Advantages: Recasting the Debate   89
    Mary C. Waters

8   The Changing Role of Race and Ethnicity in the
    Incorporation of Refugees, Immigrants, and their Children   101
    Philip Kasinitz

9   Young, Gifted, and West African: Transnational Migrants
    Growing Up in America   113
    Marilyn Halter and Violet M. Showers Johnson

10 Mexican Immigrant Legalization and Naturalization and
    Children's Economic Well-Being   129
    Frank D. Bean, Susan K. Brown, Mark A. Leach,
        and James Bachmeier

Part IV   Educating Immigrants

11 Schooling Pathways of Newcomer Immigrant Youth   151
    Carola Suárez-Orozco and Francisco X. Gaytán

12 The Importance of Raising Mexican American High School
    Graduation Rates   167
    Joel Perlmann

13 Increasing Concrete Knowledge and Community Capacity:
    How CUNY and other Institutions Can Help Reshape
    Mexican Educational Futures in New York   177
    Robert Smith

Part V   Institutions Providing Aid to Young Immigrants and Refugees

14 Tapping the Potential of Refugee Youth   191
    Robert J. Carey and Jane S. Kim
    International Rescue Committee

15 Reasons for Living and Hoping: Strangers No Longer   209
    Julianne Duncan
    United States Conference of Catholic Bishops, Migration and
        Refugee Services

16 Protection of Refugee and Migrant Children: The Role
    of Nonprofits   225
    Annie Wilson
    Lutheran Immigration and Refugee Service

| 17 | A Framework for the Care of Unaccompanied Children | 237 |

Ken Tota
Office of Refugee Resettlement, U. S. Department of Health & Human Services

Part VI Looking into the Future

| 18 | Home Country Farewell: The Withering of Immigrants' "Transnational" Ties | 253 |

Roger Waldinger

| 19 | Is There a Looming Period of Liminality for Race and Ethnicity in the United States? | 267 |

Richard Alba and Hui-shien Tsao

References	279
Contributors	303
Index	311

List of Tables

6.1	Summary of refugee admissions for FY 1975–FY 2005	77
6.2	Selected nationalities, percent in occupational categories	78
6.3	Selected nationalities, social characteristics	78
10.1	Entry Status and Legal and Naturalization Trajectory, Fathers and Mothers of 1.5 and Second Generation Mexican-Origin Respondents	135
10.2	Percent of Fathers and Mothers with Various Legal and Naturalization Statuses, 1.5 and Second Generation Mexican-Origin Respondents	137
10.3	Indicators of Human Capital and Economic Attainment by Father's and Mother's Entry Status and Legal Status at Time of Interview, 1.5 and Second Generation Mexican-Origin Respondents	138
10.4	Human Capital and Economic Attainment by Father's and Mother's Entry Status and Legal Status and Naturalization Trajectory, 1.5 and 2nd Generation Mexican-Origin Respondents	141
11.1	Correlation matrix	154
11.2	Predicting GPA—Student Centered Perspective	159
11.3	Predicting Standardized Achievement Test—Student-Centered Perspective	160
11.4	Predicting Standardized Achievement Test—School Perspective	161
12.1	Modeling the improvements in ethnic earnings that would result from improvements in ethnic educational attainment, men 25–34 in two groups	169

12.2	Male educational attainment among second generation Mexicans and native whites: actual and hypothetical	174
18.1	Cross-border engagements	259
18.2	National attachment and identity: The United States versus home country	262
18.3	U.S. political participation (naturalized citizens only)	263
19.1	The racial, ethnic, and nativity composition (%) of the best-paid occupations by age group, 2000	272
19.2	The racial, ethnic, and nativity composition (%) of the best-paid occupations by age group, 2005	273
19.3	U.S.-born, non-Hispanic white birth cohorts, counted in 2000 and projected for 2020	275

Introduction

Gerhard Sonnert and Gerald Holton

Among the darkest facts of our time, manifest in so many parts of our globe, is the never-ending, pitiful flood of refugees struggling for life. Some of the "lucky" ones among them are those who, despite the intervening horrors of their persecution (for political, ethnic, religious, or other reasons), reach that historic destination for refugees, the United States. Often traumatized and destitute, they have come and are still coming from Africa, Asia, Russia, South America, and many other regions. Refugees constitute a subgroup within a much larger set of people: immigrants. All types of immigrants face challenges inherent in the uprooting and disorientation that commonly accompany migration from one country to another, from one culture to another, and, in most cases, from one language to another. Refugees confront those challenges in perhaps their most extreme form.

The arrival of refugees and immigrants is now viewed rather critically in many quarters, although, since its foundation, the United States has been the proverbial "nation of immigrants." With the exception of Native Americans, almost all persons in this country trace their family lineage, within a small number of generations, to forebears who arrived in America as immigrants, most voluntarily though some as slaves. Large numbers of people are still coming in search of a better life, of freedom, and of economic and cultural opportunity. As modern transportation and communication technologies are advancing, the potential for immigration might even increase in the future. Immigration has thus been one of the fundamental issues that has defined America and with which American society has been struggling almost constantly and often with great intensity.

Political, economic, and scholarly attention has mostly focused on adult immigrants and refugees, even though most migration streams have contained sizeable segments of children and youths. The old hypothesis that young refugees and immigrants would assimilate quasi automatically, and without much ado, into the American mainstream (e.g., Saenger 1941), if it ever was true, is patently inadequate for the current situation. Hence, serious attention needs to be paid to young refugees and immigrants.[1] These children and youths, we believe, present for our country a special moral challenge—and even a special opportunity. That challenge, that opportunity, is to explore what help these young people need and can be given, to enable them to achieve more easily a good life and successful careers, and to let them become valuable assets to their new country. The present volume intends to make a contribution toward helping young refugees and immigrants succeed.

Here is a brief account of how this book came about. Its origins lay in our research study called "Project Second Wave," about young refugees to the United States from National Socialism. As, despite some initial difficulties experienced by this cohort, the contours of a success story, at the collective level, emerged, we also realized that we should not, and could not, simply abandon this field after the completion of the study (whose detailed results are presented in the book *What Happened to the Children Who Fled Nazi Persecution*, Palgrave Macmillan, 2006). We began to hope that our work would be more than a historical study, and that it would have some relevance for the new waves of immigrants and refugees now reaching our shores. Even though the situation today is, of course, different in many respects from that facing our study cohort, young refugees who arrived in the 1930s and 1940s, it seemed certainly worthwhile, even imperative, to ask what lessons might be gleaned from the experience of those young refugees we had studied for the work of today's aid organizations, scholars, and policy makers. These lessons, when suitably adjusted, might be useful to help the current young arrivals to achieve better outcomes, for both themselves and their new country. Even more generally, we wanted to understand the basic characteristics of the situation in which today's young immigrants find themselves—what barriers they face and what opportunities they have.

That was the mandate to which we committed ourselves. We felt that a project to carry it out should have two major features. First, we realized that this follow-up project, it if was to be useful, transcended our own area of expertise, and that of any one individual. It needed to be collaborative, bringing together the knowledge and skills of many of the leading scholars in this general area who would approach our topic with a variety of perspectives (historical, sociological, psychological, legal, economic, etc.); methods (quantitative and qualitative); and groups studied (Mexicans, Vietnamese, Russians, Africans, etc.). Second, our purpose was bridging the gap between two communities—the scholars who study immigration-related phenomena, and the humanitarian organizations and institutions that provide concrete help to immigrants and refugees.

On the one side, immigration has, of course, been a major and long-lasting area of scholarly investigation. The scholars who contributed to this volume have been at the forefront of these research efforts. On the other side, there are numerous organizations that have helped immigrants and refugees. In a strong and enduring American tradition of citizens' activism, described already by Alexis de Tocqueville, American nongovernmental organizations early became involved in aiding immigrants and refugees to settle into their new country and to become productive members of American society. They collectively have had a long tradition of playing a crucial role as enablers of the American Dream. The federal government has more recently become a major actor in this arena, notably since the creation of the U.S. Office of Refugee Resettlement as a consequence of the Refugee Act of 1980.

Thus, we wanted to initiate a dialogue between these two rather distinct communities of practitioners and scholars. This is what set our project apart from a purely academic endeavor. A conceptual model of research called Jeffersonian Research guided our approach. We proposed this model as a complement to the ubiquitous—but, in our opinion, flawed—dichotomy of basic and applied science (Holton and Sonnert 1999, Branscomb, Holton and Sonnert 2002, Sonnert and Holton 2002). At the abstract level, Jeffersonian Research is *basic* science in the social interest; it places basic research in an area of basic scientific ignorance that seems to lie at the heart of a social problem. We saw the opportunity that a dialogue between aid organizations and scholars in the area of immigration might evolve into an example of the fruitful application of the Jeffersonian Research concept. The knowledge accumulated by the scholars might help the aid organizations to fulfill their mission, while in turn the scholars might be motivated to situate future research in areas in which the organizations need more information.

We therefore organized a conference at the Russell Sage Foundation headquarters in New York City in November 2006 and brought together representatives of both major refugee and immigrant aid organizations and the U.S. government, as well as leading scholars in the area of migration studies. The revised and updated contributions are presented in this volume. They are arranged in six parts that are connected in an approximately chronological pattern. The *first* part provides a broad historical perspective on the reception and the life paths of young immigrants and refugees in the United States, the *second* focuses on the historical case of the "Second Wave" young refugees, the *third* turns to more recent and current immigration streams, the *fourth* highlights one crucial determinant of the life chances of young immigrants and refugees—education, the *fifth* presents the concrete work of aid organizations, and the *sixth* looks to future developments. At the beginning of each part, a short introduction summarizes the individual chapters in that section and explains how these chapters are related to the broader theme of the book.

Together, these chapters provide a rich portrait by leading scholars of the fate of young immigrants and refugees in the United States, as well as a description of concrete aid efforts undertaken by major private aid organizations and the U.S. government. Many insights, some of them counterintuitive, can be garnered from these pages. This multifaceted collection of research findings, insights, and aid practices is, we hope, at least a beginning, and will stimulate further contact between the practitioner and the scholarly communities in the area of refugees and immigration that might eventually bring about a Jeffersonian-type research program.

From the various presentations, we perceived needs among the aid organizations that might be addressed with the help of scholars. The organizations' typical mode of activity is in the realm of emergency response. Humanitarian disasters happen all too frequently around the globe, and immediate help is of the essence. A short-term focus, thus, is not surprising. Yet, the issue of young refugees and immigrants also has clear long-term implications, and it may be highly beneficial to aid organizations if long-range perspectives offered by scholars are also taken into consideration.

There seemed to be a need expressed not only for evaluation research—helping the organizations to analyze which of their programs and activities worked well, and which did not work so well—but, more fundamentally, a need for understanding how "success" could reasonably be defined in the first place. And then there are the potentially significant benefits to be derived from both the accumulated scholarly knowledge about long-term processes and their determinants—about how refugees and immigrants do (or do not, even over generations) integrate into America after their arrival—and the new research in this area. This has been the province of social scientists for whom large-scale and long-term processes of assimilation and social mobility have been of major interest.

In general, scholarly research of the kind presented here sharpens awareness of the unintended consequences of policies and interventions. The analyses of the fate of young refugees and immigrants furthermore demonstrate the value of a balanced conceptual approach, according to which social outcomes arise out of an interplay of societal opportunity structures and individuals' will and ability to take advantage of them. Deviating from this balanced approach on both extremes may be pernicious. Merely inspirational speeches to young immigrants, that all it takes for them to succeed is to "work hard," and similar exhortations that ignore the societal structures the young immigrants and refugees have to deal with, are naïve at best. On the other hand, focusing solely on social structures and on how to transform them while ignoring the young immigrant or refugee as an individual agent (for fear of blaming the victim) seems equally problematic. Adverse structures may not be insurmountable, and even profitable opportunity structures will have no effect if not taken advantage of.

America is certainly the land of opportunities for immigrants—but it is just as certainly the land of what one might consider a generalized Matthew effect (which Robert K. Merton originally postulated in the context of the sociology of science), driving up the variability of outcomes. That is, those who have advantageous starting positions—in terms of cultural or social capital, social networks and support structures, and personal drive—have the chance to do spectacularly well, whereas those who do not are in danger of being left behind. An instantiation of this very general principle in the area of young immigrants and refugees is the paradox noted or implied in several papers: people who are particularly adept at availing themselves of aid may not be the ones who need it the most, but may tend to be people who might have done well anyway. Yet, organizations should not feel discouraged from engaging in aid activities that might not create a superficially impressive record, but that can be very valuable indeed.

We thank the Russell Sage Foundation and its president, Eric Wanner, for generously supporting the conference. We also gratefully acknowledge the crucial financial assistance by a donor who chose to remain anonymous. We thank all the contributors for participating in this collaborative project and for sharing their expertise, insight, and wisdom. Our hope is that this book will be useful in helping young immigrants and refugees succeed.

Note

1. Among the laudable pioneering efforts in this area are those by Kasinitz et al. (2004), McGill (2005), Min (2002), Portes (1996), Portes and Rumbaut (2001), Robben and Suárez-Orozco (2000), Rumbaut and Portes (2001), Suárez-Orozco and Suárez-Orozco (1995, 2001), and Zhou and Bankston (1998).

Part I

Setting the Stage: The Historical Framework

*F*or a deeper understanding of the current situation of young immigrants and refugees coming to the United States, and of the challenges and opportunities that present themselves to those who wish to help them, a useful starting point is a survey of the larger historical trends that will locate current issues in their historical context.

The two chapters in this section provide such a historical framework. Nancy Foner's chapter addresses the history of early twentieth-century immigration. She examines two major contingents of the immigration wave from around 1880 to the mid-1920s—Eastern European Jews and Southern Italians—and charts the subsequent developments of these immigrant communities.

Reed Ueda's contribution focuses on the controversial public debate that accompanied those earlier immigration waves, showing that key arguments made in the current debate have intellectual roots that reach back at least a century. The chapter also points out that the immigration-inflicted doom predicted by the earlier anti-immigration rhetoric did not materialize.

Questions of Success: Lessons from the Last Great Immigration

Nancy Foner

Understanding the factors that help—or hinder—the success of immigrant children is a pressing issue in contemporary American society as the number and proportion of foreign born and their children continue to grow. Not surprisingly, many scholars and popular commentators writing about this topic set their analyses against the backdrop of the last great immigration of the turn of the twentieth century when the proportion (though not absolute number) of immigrants in the United States was even larger than it is today. In 1910, 35 percent of U.S. residents, or 56 million people, were born abroad or were the U.S.-born children of immigrants. In 2006, the number was almost 70 million, although given the much larger U.S. population, the share of immigrants and U.S.-born children of immigrants was close to a fourth.

In the quest to contribute to what might be called a "sociology of success," this chapter offers a historical approach, looking at the children of immigrants in the earlier era.[1] From around 1880 to the mid-1920s, more than 23 million immigrants, mostly southern, eastern, and central Europeans, came to the United States. The focus here is on eastern European Jews and southern Italians, the two largest ethnic contingents of the time. I draw heavily on research on New York City, where one out of seven of the nation's immigrants—and more than a third of the Russian Jewish immigrants and a quarter of the Italians—lived in 1910.[2]

This chapter centers on two principal questions. Each is concerned with the past, yet each also has implications for the present. First, was the progress of the children of immigrants in the earlier influx as rapid and problem free as many accounts assume? The answer, I will argue, is most decidedly no. Exclusively upbeat portrayals of the past fail to capture the complexity of second generation mobility paths in previous periods, including barriers encountered along the way. Moreover, by putting such a positive gloss on the past, these portrayals create unrealistic expectations for—and reinforce overly pessimistic predictions about—today's second generation. The second question relates to the *relative* success of particular groups: specifically, why were the children of eastern European Jewish immigrants more successful educationally and occupationally than the children of Italians? Again, the analysis has significance for the present. Accounting for differences among ethnic or national-origin groups is a key, some would even say *the* key, question for research on the children of contemporary immigrants, most of whom come from Latin America, Asia, and the Caribbean (Zhou 2001: 301). Many of the factors in the Jewish success story—cultural as well as structural—provide insights into understanding why some contemporary immigrant groups have disproportionate numbers of high achievers. At the same time, new dynamics have also emerged as important.

Second Generation Progress in the Past: How Easy and Rapid Was It?

The rags-to-riches story is a basic part of the American dream. European immigrants, the story goes, sweated and struggled when they got here, but their children were able to "make it" in America. If "we" did so well, say many Americans whose parents and grandparents came a century ago, why can't the children of the newest arrivals? Many academic commentators also hark back to the success of the children of earlier immigrants, mainly focusing on the positive aspects of their experiences and arguing that today's second generation faces far more uncertain and contingent prospects (for a well-known exception, see Perlmann and Waldinger 1999).[3]

In fact, the ascent up the socioeconomic ladder was more difficult for yesterday's second generation than is often remembered by their descendants or assumed in the scholarly literature. To be sure, the story of yesterday's second generation is overall one of progress and advancement. But the climb up was often slow and gradual rather than a matter of giant leaps forward, and some of the second generation suffered painful setbacks and obstacles along the way.

"My son the doctor" may have been a cherished phrase of Jewish immigrant parents but "my grandson the doctor" is more accurate. Indeed, for turn-of-the-twentieth-century European immigrants, the leap into the professions was generally a third or fourth generation phenomenon. Second generation Europeans usually made relatively modest moves up the socioeconomic ladder when compared to their parents—as Joel Perlmann (2005) puts it, typically they did appreciably, rather than vastly, better than their parents—and some of course did not move up at all, but stayed at the same level.

In 1950, in New York City, a quarter century after the massive influx from southern and eastern Europe had ended, the majority of the Italian and the Jewish second generation were clerks, skilled workers, and small business owners. Only a small proportion of second generation Jews—and an even smaller proportion of second generation Italians—were then in the professions. Blue-collar work continued to be the mainstay of the Italian second generation among males and females alike; Jews were more likely to be found in clerical and sales jobs and as managers and proprietors. It was not until the 1950s that (mostly third generation) Jews really began a mass program of college education; for Italians it would not be until a decade or two later. It took almost 100 years from the time the Italians' mass immigration began before it became clear that their descendants as a group would make it, educationally and occupationally, into the American mainstream.

The children of eastern and southern European immigrants also faced strong exclusionary forces in their drive to succeed. Those who came of age in the early twentieth century did so at a time when Jews and Italians were seen as racially different from—and inferior to—people with origins in northern and western Europe. They were legally white, and thus, unlike Asians, permitted to naturalize as U.S. citizens and not subject to anti-miscegenation laws that existed in many states. But, at the same, they were seen as "racially distinct from other whites"—those of Nordic or Anglo-Saxon background who were viewed as the genetic fundament of the American stock (Jacobson 1998, see Ueda, in this volume). As George Fredrickson has written: "Whether they were a genetically inferior subcategory of whites or an 'inbetween people' who did not quality as white, they were considered racially 'other' and different in ways that went beyond acquired cultural characteristics" (Foner and Fredrickson 2004, see Barrett and Roediger 1997, Jacobson 1998, Foner 2005). Jews and Italians were believed to have distinctive biological features, mental abilities, and innate character traits that marked them as inferior to northern and western Europeans. And they were thought to be physically identifiable—large noses often noted, for example, in the case of Jews, and "swarthy" skin in the case of Italians.

As members of the eastern and southern European second generation sought to advance, they often faced discrimination and exclusionary policies by largely Protestant middle and upper-middle classes who attempted to preserve their privileges. Jews were frequently a target, mainly because of their relatively early educational gains. During the 1920s, when second generation eastern European Jews acquired the educational credentials to gain entrance into Ivy League universities in large numbers, efforts were made to keep them out. Harvard developed limits on the number of Jewish students admitted, and in one form or another, many other elite colleges followed its lead. From the 1920s onward, many Jews were denied admission to professional schools; they were not hired by elite law forms, could not gain faculty positions at universities, and were systematically barred from elite social clubs (see Dinnerstein 1994: 85–92). (It was not until 1946 that the New York City Council passed a resolution threatening the tax exempt status of nonsectarian colleges and universities that used racial or religious criteria in selecting students; in 1948, New York State followed with an antidiscrimination statute.) In the Depression years, New York City's largest employers—including public utilities, banks, insurance companies, and home offices of major corporations—rarely hired Jews. Non-Jewish law firms rarely took on Jewish attorneys; medical and dental offices did not welcome Jewish graduates (Wenger 1996: 22–24). Jewish professionals—accountants, lawyers, doctors, and dentists—fell back on a pattern of self-employment, setting up their own practices and catering to a largely Jewish clientele. In the 1930s, Jews became a growing presence in the civil service after Mayor LaGuardia implemented a merit system for hiring and promotion.

Both Jews and Italians faced obstacles in moving to the most desirable neighborhoods. Restrictive covenants, clauses in real estate titles that limited the sale or transfer of property to members of certain groups, frequently excluded Jews from buying homes in privileged neighborhoods—and these clauses were not outlawed by the U.S. Supreme Court until 1948.

Sheer economic circumstances were also impediments. American economic ascendancy, the expansion of higher education, suburbanization, and generous government assistance to veterans—these specific historical conditions in the post-World War II years are usually mentioned as facilitating the upward mobility of the eastern and southern European second generation. Often overlooked is that a substantial part of this generation was born too early—before 1910, say—to benefit much from these conditions. It is also well to remember the devastations of the Great Depression. The Depression affected all but a small number of the second generation (those who were born after it ended), and for most it meant economic and social dislocations, even if these were temporary. Earlier cohorts of the Jewish and Italian second generation (and many who arrived in the United States as young children) came of age during the 1930s and, as a result, suffered diminished opportunities, which in many cases derailed and disrupted school and work careers.

Why Were Jews More Successful than Italians?

Italians and Jews were the two large immigrant groups of their time, both poor and despised when they came to America. Yet the mobility patterns of their children were markedly different. Jews were an unusually successful group, whose offspring reached middle-class status in fewer decades and in fewer generations than did Italians—and why this was the case has long preoccupied historians and social scientists (Perlmann 2000: 103).

The greater achievements of the children of Jewish immigrants in school and in the workplace were evident from the very beginning. In New York City in the first decade of the twentieth century, Russian Jewish children surpassed Italians on every measure: a smaller proportion of Russian Jews were truant, were left back, or dropped out of school before it was legally permissible, and a much higher proportion went to and completed high school and college (Foner 2000: 194–196). The Jewish advantage in education and occupation persisted throughout the century; in 1950, second generation Jews were three times as likely as second generation Italians in the New York area to be in the professions or managerial occupations, while second generation Italian men were twice as likely (and second generation Italian women more than four times as likely) as their Jewish counterparts to be operatives (Glazer and Moynihan 1970: 322–323).

The social and economic background of the two groups clearly plays a major, indeed *the* major, role in accounting for these differences. But cultural background, I would argue, is also part of the story.

When it comes to structural or social and economic background factors, eastern European Jews, for one thing, had an occupational head start. Compared with southern Italians, they were more urban (or at least had experience in small towns) and arrived with higher levels of skills—and transferable skills—which gave them a leg up in entering New York's economy. Jews' concentration in manufacturing occupations in eastern Europe (including manufacturing handicrafts) was a source of advantage when they came to America, compared with the Italians' background in rural agriculture and, so too, Joel Perlmann (2000) argues, was the experience of large numbers of Jewish immigrants in petty trade and running small businesses.

Because the Jewish immigrant population was, from the start, better off economically than the Italian, Jewish parents could afford to keep their children in school more regularly and for longer. The poorer, less-skilled southern Italians were more in need of their children's labor to help in the family. This was true for girls as well as boys. Indeed, Miriam Cohen (1992) argues that because more Jewish mothers could afford to remain out of the paid workforce, fewer daughters had to leave school to assist in either homework tasks or child care.

Although eastern European Jews who came to this country were not, as a group, highly educated, the fact that they were more literate than the Italians was also a help. Only about a quarter of Jewish immigrants over the age of 14 entering the United States between 1899 and 1910—compared with more than half of the southern Italians—told immigration officials they could not read or write. If Jewish children were more likely to have literate parents, they, too, often arrived with a reading and writing knowledge of one language, making it easier to learn to read and write English than it was for southern Italian immigrant children, who generally arrived with no such skills (Cohen 1992).

What about culture? This has been, to put it mildly, a contentious issue in the literature, with sociologists particularly hostile to cultural explanations. Viewing culture as a reason for the failure of a group to achieve social mobility is seen as "blaming the victim." Cultural explanations, as Orlando Patterson (2000: 204) has written, are often "employed by reactionary analysts and public figures who attribute the problems of the poor to their 'values' and thereby wash their hands, the hands of the government and the taxpayers of any responsibility for their alleviation... Culture as explanation languishes in intellectual exile partly because of guilt by association." There is the assumption that culture is immutable—and that nothing can be done about it—thereby providing a rationale for policy makers to withhold their support from policies that would ameliorate the economic and social conditions of the poor.

Given this ideological baggage, it is no wonder that scholars often seek to avoid, and thereby overlook, cultural factors. There is a need to move beyond "politicized dichotomies of structure vs. culture" and, at the same time, to take culture seriously (Lamont 1999, Waters 1999: 63–64). Indeed, by shunning the concept of culture, there is a risk, as Hans Vermeulen (2000: 2) has argued, of obscuring important issues, however controversial and sensitive these may be. Culture—and by this I refer to a repertoire of socially transmitted symbols, values, and ideas and patterned ways of life associated with them—can make a difference. Its role may be modest, but it ought at the very least to be considered as one of the possible—and complex—interacting factors accounting for ethnic differences (ibid.: 11).

Cultural outlooks, attitudes, and values are of course rooted in social, economic, and political forces; they also, invariably, change over time in response to new conditions and opportunities. Yet values do not cease to be values once they can be related to socioeconomic conditions that precede them (Vermeulen and Venema 2000: 126). Moreover, cultural patterns that were shaped in premigration settings can have a survival power and continue to operate (even if in modified form) in ways that influence behavior after migration. Culture, in short, is not simply reducible to structure—and can have some autonomy and a life of its own.

One reason that eastern European Jews did so much better academically than Italians was that their immigrant parents placed a high value on education. Jews came from a culture in which religious scholars were the most honored individuals in the community; families proudly supported the learned men who spent their days reading the Talmud and disputing its meaning. In stark contrast, southern Italians' cultural heritage made them less oriented to book learning and less sensitive to the economic advantages of schooling (Perlmann 1988: 115, see also Alba 1985: 59–62). By some accounts, southern Italians were antagonistic to schools and intellectuals in their home country. At the least, they were skeptical about the value of education. In peasant communities in southern Italy, formal schooling was viewed as largely irrelevant for their lives. There was "no obvious need for more than a trifling amount of formal education"; the practical skills needed for survival could be taught at home or through apprenticeship. Knowledge beyond the everyday requirements was a luxury meant for the "better classes." Time spent in school simply took children away from work (Covello 1967).

That Italian cultural heritage mattered comes out in Perlmann's detailed study of Providence, Rhode Island: Italian children from middle-class homes left school earlier than Russian Jewish children from working-class homes and earlier than other children from working-class homes as well. Families had to make choices about resources and needs, he suggests, and these were determined partly by material constraints but also partly by cultural factors, including views on the value of schooling and the proper role of adolescents in terms of contributing to the family economy (Perlmann 1988: 216–217). Premigration cultural attributes, he writes, cannot be dismissed or even treated as afterthoughts, but rather constitute an important part of the explanation for group differences in behavior (ibid.: 216).

Different attitudes may have affected the school performance of Jewish and Italian children in another way. Jewish pupils in New York at the turn of the twentieth century, with their positive attitude to schooling, were said to be the "delight of their teachers"; Italian children were "more or less difficult to discipline and irresponsible." Many Jewish students strove to live up to their teachers' high expectations, whereas Italian children had to struggle against negative stereotypes (Berrol 1974: 36).

Does Culture Play a Role Today?

As in the past, contemporary immigrants' social and economic background—the human capital they bring with them—is the most significant factor in explaining their children's success. Not surprisingly, immigrant groups with high proportions of highly skilled and educated members—Asian Indians, Koreans, and Soviet Jews, for example—also have high proportions of educationally and occupationally successful children. Highly educated immigrant parents, as many studies show, have higher educational expectations for their children and provide family environments more conducive to educational attainment. If their children started school in the home country, they typically attended excellent and rigorous institutions. If the parents managed to obtain good jobs in the mainstream economy in the United States—which many highly educated immigrants have done—this has provided the income to live in better neighborhoods with better schools. Even if they have experienced significant downward occupational mobility in this country, well-educated parents are usually more sophisticated about the way the educational system works here and have an easier time, and more confidence, in navigating its complexities—and steering their children into good schools—than those with less education.

Yet critical as socioeconomic background factors are, they are not sufficient to explain the achievements of Asian youth whose parents have only modest levels of education and low incomes—and, more generally, why a number of national studies show students of Asian origin outperforming those in all other racial/ethnic groups (including non-Hispanic whites), even after taking into account such factors as family income, household composition, and parental education (Foner 2000).

One set of explanations emphasizes the dense social networks in many Asian communities (Zhou and Bankston 1998: 148, Portes and Rumbaut 2001: 260–261). Still, if strong ethnic communities are important, the question remains as to why Asian ethnic communities are such powerful forces for success, while others, like Mexican communities, are not (Lopez 2008, see Suárez-Orozco and Gaytán, in this volume). A recent study of the New York second generation argues that at least in the Chinese case, the high degree of connection between better- and worse-off immigrants in New York City's Chinese communities is a big factor in that it gives working-class Chinese immigrants access to information about how the educational system works and how to get their children into better schools (Kasinitz et al. 2008).

It also seems likely that premigration cultural legacies (whatever their origins) play a role in the greater success of children in many Asian immigrant groups. I am not simply talking about cultural capital of the sort that Gerhard Sonnert and Gerald Holton (2006) contend is implicated in the extraordinary success of German Jewish refugees who fled to the United States as children or youths in the World War II era. By cultural capital they have in mind, following Pierre Bourdieu, long-lasting "dispositions of the mind and body... imbibed in early years from families and the surrounding environment." German Jewish refugees, Sonnert and Holton argue, benefited from their earlier socialization emphasizing "high culture" (classical music, art, and literature), broad intellectual interests, as well as discipline and intense hard work. In this case, cultural capital is certainly class linked—related to the fact that more than 70 percent of the participants in Sonnert and Holton's survey were from an upper-middle-class (or in a few cases upper class) background in Germany.

When it comes to east Asians in the contemporary era, the relevant cultural messages and values are less clearly related to social class in the country of origin. It seems to be a question of widely diffused premigration values that rate education extremely highly and support behaviors that facilitate educational achievement (Zhou 1997a: 194, Min 2006: 86). Confucian teaching, it is said, not only puts a high value on education but also emphasizes discipline, family unity, and obedience to authority, all of which contribute to academic success. Children who do poorly in school bring shame to Chinese families; those who do well bring honor. If their children study long hours, Chinese immigrant parents believe they can get As—and they push their children to "work at least twice as hard as their American peers" (Zhou 1997a: 193–194, see Min 1998 on Koreans).

It may be, too, that the traditional respect for scholars and teachers in Chinese culture contributes to positive student-teacher interactions in the United States (Sung 1987). Moreover, the cultural beliefs of the teachers also operate in that the model minority stereotype held by many of them benefits Chinese and other Asian students (Kasinitz et al. 2008: 169). In a timeworn pattern, high teacher expectations help bring about high student performance.

Min Zhou (1997a) raises another factor that deserves further exploration: channeling funds for purposes that encourage and support educational success. Chinese immigrant parents, she writes, value thrift and denounce consumption of name-brand clothes and other "too American" luxuries. At the same time, they do not hesitate to spend money on books, after-school programs, Chinese lessons, private tutors, music lessons, and other educationally oriented activities.

The cultural background of Korean and Chinese immigrants operates in still another way. They have imported after-school institutions that, in the New York City setting, prepare their children for entrance examinations to the city's specialized high schools and later on elite universities. These private academies are a tradition in Asia, where competition for high-quality education and to get into top universities is fierce. Direct involvement in or exposure to institutionalized supplementary education in China and Korea, as Zhou and Susan Kim (2006) note, is part of the cultural repertoire that both middle-class and working-class Chinese and Koreans carry with them when they migrate to the United States. In New York's Chinatown, where tutor services and test preparation programs are readily available, "school after school," according to Zhou (1997a: 202), has become an accepted norm (see Zhou 2009 on the rapid growth of Chinese after-schools in southern California). According to one survey, a fifth of Korean junior and high school students in New York City were taking lessons after school, either in a private institution or with a private tutor (Dunn 1995, Macdonald 1995).

Quite apart from these "cram schools," dozens of ethnic language schools exist in New York City's Chinese community, and by one estimate, there are more than 100 Korean language schools in the New York area run by Korean churches that teach children the Korean language, history, and culture (Min 2001: 187). However successful ethnic language schools are in inculcating homeland cultural traditions and language skills, from a mobility perspective what is most important is that they encourage and reinforce habits of study as well as parental values about the importance of schooling. Moreover, they do so in a setting that is controlled within the ethnic community—rather than provided by benevolent outsiders—thereby reinforcing a sense of identity and cultural pride (cf. Lopez 2008).

A final way that culture operates should be noted. In their study of the New York second generation, Philip Kasinitz and his colleagues suggest that the success of the children of Chinese immigrants is partly due to the pattern of family obligations that keeps working-class mothers and fathers together—thereby boosting the family income—and encourages second generation Chinese to put off marriage and childbearing until they have finished school and established themselves in their careers (Kasinitz et al. 2008).

New Factors in the Contemporary Era

If today, as in the past, socioeconomic and cultural background factors give some immigrant groups advantages over others in facilitating the success of their children, a number of new elements are also involved.

One concerns the role of strong networks and premigration cultures in the ethnic community. As I have already noted, analyses show that remaining engaged with the ethnic community and ethnic culture can help the children of today's immigrants get ahead. At the same time, acquiring the habits and attitudes of teenagers in the contemporary American mainstream—as well as those of many inner-city native-born minority youth—is seen as having the potential to undermine positive immigrant values. In his chapter, Reed Ueda argues that these concerns have a familiar ring. He notes that many social reformers 100 years ago also worried that modern urban popular culture was potentially harmful and would lead some children of immigrants astray as they rejected foreign ways and parental discipline. Yet in earlier times, voices skeptical of the promise of assimilation for the children of European immigrants were in a minority among intellectuals and social scientists and in the immigrant communities themselves (Foner and Kasinitz 2007). There were strong pressures for children to acquire American habits and ways. In the early twentieth century, educators were adamant that the children of immigrants had to Americanize and shed much of their immigrant culture to get ahead, and they saw their role as weaning immigrants away from their ethnic heritage (Foner 2000: 206–214). Indeed, a concern was that among some groups, like Italians, the cultural values and close families in tightly knit immigrant communities were *preventing* the second generation from making educational progress. Today, immigrant cultural beliefs and family structures are often seen as giving them an advantage.

Race also plays a different role today, which is striking in New York with its large number of east Asian as well as black and Hispanic immigrants. In the past, race was basically irrelevant in explaining why Jews did better academically than Italians. Both groups were at the bottom of New York City's ethnic pecking order, considered to be inferior white races. Today, the way east Asian—as opposed to black or Hispanic—immigrants fit into New York's racial hierarchy makes a difference in the opportunities they can provide their children. Because they are not black, east Asian (and white) immigrants face less discrimination in finding a place to live and, in turn, a school for their children—which translates into greater access to heavily white neighborhoods with good public schools. Moreover, even if they attend school with many native-born blacks and Latinos, the children of east Asian (and white) immigrants do not, like the children of black and Latino immigrants, feel a bond of race with native minorities or share the experience of intense levels of prejudice and discrimination or the same kind of negative racial stereotyping from teachers (also see Halter and Johnson, in this volume, on West African immigrants). One result is that the children of east Asian immigrants are less likely to become part of a peer culture found among some disaffected inner-city black and Latino youth that emphasizes racial solidarity, disruptive and dangerous behavior, and often hostility to school rules and authorities.

Conclusion

Studying the past, it is often said, can illuminate the present, and this analysis of the pathways to—and factors involved in—success among the children of immigrants in the last great wave has implications for understanding these dynamics in the present period. The analysis thus not only has relevance for the scholarship on immigration, but, I believe, may also be useful in informing the work of organizations assisting newcomers today—organizations that, in some cases, were founded to help earlier European arrivals.

Much depends of course on how we understand history, and one lesson to be learned from the experiences of the earlier second generation is to be wary of romanticizing the past. A common tendency is to emphasize only the positives in looking back to the European second generation's mobility paths. As I have shown, progress up the socioeconomic ladder was not problem free for the children of southern and eastern European immigrants who arrived in the United States a century ago. A too-rosy view that glosses over the obstacles and barriers in the past suggests that the present second generation is unique in facing incorporation difficulties. This is misleading. An awareness of the complexities involved in the path to mobility in the past—including the stumbling blocks along the way and the often modest moves involved—provides a more realistic basis for comparison and for theorizing about the prospects of the second generation today.

There are also lessons for the present to be learned from comparing the different trajectories of the children of Jewish and Italian immigrants in the last century. To be sure, a number of factors involved today—such as race—did not operate, or were less important, in explaining the greater achievements of second generation Jews. Yet revisiting the past reaffirms that then, as now, social and economic background factors—above all, immigrants' human capital—are of primary significance in explaining differing rates of success among the second generation in various national-origin groups. Going back to analyze why the children of Jewish immigrants did so much better than those of Italians also strongly suggests that cultural background comes into the equation. Much as sociologists are often loathe to assign much, if any, weight to culture, the final lesson that bears repeating is that cultural factors should not be rejected out of hand. They ought to be given serious consideration in trying to make sense of patterns of success among ethnic groups—and this is just as true today as it is for the past.

Notes

1. The category "children of immigrants" includes those born in the United States as well as those who arrived as young children (often referred to as the 1.5 generation).
2. Jews were about one of seven, and Italians more than a third, of the immigrants arriving in the United States between 1899 and 1924 who did not go back (Perlmann 2005).
3. This section of the chapter on second generation progress in the past draws heavily on Foner and Alba (2006).

2

Conceptual Frameworks for Obstacles to Immigrant Advancement: Perceiving Immigrants in the Context of Societal Change

Reed Ueda

The following is a heuristic reflection on one of the specific issues for this book: what have been and/or are now the major obstacles to immigrant advancement? As a historian, I have been interested in the background of social thought about this topic and the possibility of describing its conceptual framework. This chapter will focus on the decades surrounding the turn of the last century when many critical ideas about the "destiny" of young immigrants were generated. Its purpose is to create an opportunity to discuss possible implications of historical perspective for today's "immigration mind-set" in the fields of journalism and politics, which have the potential to affect the policies of organizations working in the field of immigrant aid.

Perceiving the Marginality and Vulnerability of Young Immigrants

Influential groups of public commentators in the late nineteenth and early twentieth centuries grew concerned with the existence of limited economic and cultural channels for young immigrants and linked their life prospects to what were perceived as unprecedented macro-structural changes: the decline of social mobility, the formation of a transnational transient underclass, the disintegration of national identity, the rise of ethno-racial hierarchies, a debilitating and exploitative mass culture, ethnic and religious separatism, and a politics corrupted by narrow group interests. Many social reformers and members of the middle-class public believed that young immigrants were becoming vulnerable to harmful pressures from these societal shifts and were facing limited or downward social mobility, family and community disorganization, cultural dysfunction, and a lack of incorporation in civic democracy. Young immigrants were perceived as a particularly vulnerable group because of what was believed to be their distinctive condition of marginality, formed from being concurrently young and foreign. Their special marginality involved being exposed or suspended, without a fully developed and functional personal-life framework, between their parents' tradition-based ethno-cultural world and a new peer-centered urban culture and society.

Educators and social reformers worried about intra-family generational conflict and loss of control over children as a problem for the adjustment of young immigrants. Jane Addams witnessed how modern urban culture pitted children acculturating to mass consumerism against parents upholding moral thrift traditions. Addams claimed, "There are many convincing illustrations" that "parental harshness" that "concedes no time or money for pleasures" has led to "juvenile delinquency." She noted, "Four fifths of the children brought into the Juvenile Court in Chicago are the children of foreigners.... Many of these children have come to grief through their premature fling into city life, having thrown off parental control as they have impatiently discarded foreign ways."[1]

Addams responded to this social problem in her settlement at Hull House where she tried to build bridges for cultural communication between parents and children. Seeing that the children of immigrants affected by an Americanized peer culture were alienated from their parents, she established a labor program at Hull House where mothers and fathers could exhibit their impressive skills in traditional crafts and arts. Addams recounted how a young Italian girl's embarrassment of her mother was turned into affectionate pride and respect when she witnessed her demonstrating spinning to an admiring audience at Hull House.

Reformers and ethnic traditionalists were frightened by an image of cultural clash between parents and children that could lead to downward assimilation of Americanizing youth into an urban mass culture based on commercialism and consumerism of entertainment. Historian John F. Kasson described how reformers issued "critiques against many of the recreations of the new mass culture" and "felt themselves to be losing control over the activities and values of the lower classes and even over their own young....":

> [These recreations] were all distrusted not only because of their commercial ends but also because of what middle-class authorities perceived as the regressive character of their activities. All of these entertainments depended on a lessened self-possession, a surrender of intellect and emotions to powerfully influential forces under the control of those who promised only pleasure and pursued only profit. (Later critics would denounce rock concerts... and hallucinogenic drugs—in similar terms.)[2]

It was feared that the desire to access popular urban culture led young immigrants to reject "foreign ways" and, as a result, discipline and authority, as well as the parents who stood for them. Housing reformer Jacob Riis's depiction of the traps awaiting young immigrants might be compared to a crude version of segmented assimilation. He drew a vivid picture of young immigrants from "disorganized" families who were vulnerable to criminality: boys were recruited into gangs and girls into prostitution.[3]

Perceiving Macroeconomic Shifts as a Limiting Context for Life Prospects

In the decades around the turn of the twentieth century, commentary on the futures of young immigrants in the United States was inseparable from an acute concern that unprecedented shifts in macroeconomic structures were distorting American social and institutional development. Many expressions occurred, on different cultural levels, of a fear that these large-scale changes would lead to the creation of a permanent underclass. A Russell Sage study characterized a poor immigrant neighborhood in Manhattan as entrapping and immobilizing its residents, likening it to "a spider's web," for of "those who come to it very few... ever leave it."[4] Dystopian science fiction depicted nightmarish versions of a permanent and antisocial underclass—the "canaille" of Ignatius Donnelly's *Caesar's Column*, and the "people of the abyss" of Jack London's *The Iron Heel*.[5] President Grover Cleveland described ordinary Americans being ground down into an underclass by the irresistible force of national industrialization. He lamented that "the wealth and luxury of our cities" was "mingled with poverty and wretchedness" of "the toiling poor." Remote and all-powerful "trusts, combinations, and monopolies"—"the people's masters"—were creating a juggernaut economy, "while the citizen is struggling far in the rear, or is trampled to death beneath an iron heel."[6]

One of the leading economists of the late nineteenth century, Francis A. Walker of MIT, argued that the occupational structure was separating into isolated "industrial layers superimposed on one another." According to historian Joseph F. Kett, Walker "ridiculed the idea that each rising generation produced a 'disposable fund of workers' who, 'eagerly watching the prospect of industry in its several branches,' silently and swiftly moved into openings." Walker instead claimed that social mobility was an illusion: "The man who is brought up to be an ordinary carpenter, mason, or smith, may go to any of these callings..." but he cannot compete "in those higher departments of skilled labor for which a more elaborate education and larger training are necessary...." Walker appeared to be conjecturing on an early prototype of a dual labor market or split labor market model, and he was not the only labor expert doing so. As Kett pointed out, "A consensus was growing among observers of the economic scene at the end of the nineteenth century.... Observers complained that youth [without higher levels and grades of education] were being forced into 'dead-end' jobs...."[7]

By 1890, fears arose that the "de-skilling" of employment—in which skilled occupations were replaced by semiskilled or unskilled jobs—was eliminating fields with career potential and thwarting the intergenerational transmission of occupational status. A perception became widespread that industrial mass manufacturing based on technological innovation was a homogenizing and degrading force that reduced workers to low-skilled and low-paid operatives. Exacerbating this worry was the tendency of young immigrants from countries such as Italy, Greece, and Poland to become "birds of passage," forming a core group of a transnational labor force that was recycled between countries of origin and the United States. As transient workers, they took low-skilled and low-paying jobs and formed a type of shadowy, alien underclass "on the move."[8]

Other factors also shaped the perception of limited or downward mobility in immigrant America at the end of the nineteenth century. An industrial working class that owned no property and only possessed labor power as its sole economic asset was growing rapidly. This phenomenon formed a disturbing contrast with the earlier farming- and artisan-dominated economy, when proportionately more people had landed property or stood to inherit it. Furthermore, the growth of poverty appeared to be accelerating. The depression of 1893–1894 produced the most severe industrial unemployment up to that point in national history. In 1904 social worker Robert Hunter published *Poverty*, which attained wide circulation, and demonstrated that in 1900 more than 10,000,000 persons out of 76,000,000 in the United States lived below the poverty line.[9] Finally, leading experts on labor issues grew concerned that immigrants were reducing job opportunities for natives and depicted the employment structure created by industrialization as a zero-sum system. Jeremiah W. Jenks, the technical director of the U.S. Immigration Commission from 1907 to 1910, was connected with this view.[10]

Perceiving Diversity as a Limiting Context for Incompatible and Transnational Groups

The specter of "two Americas" was evoked in 1890 by Jacob Riis's *How the Other Half Lives,* whose title indicated that American society was being polarized. The shift in immigration from countries in northern and western Europe after 1895 to a flow chiefly from southern and eastern Europe stimulated the perception of deepening social division and growing separation of middle-class and working-class worlds. Anxious native-born Americans of British and Protestant origins worried that this shift introduced completely unprecedented and unassimilable ethno-racial groups into American society.

Nativist perceptions of unassimilable immigrant groups were reinforced by the influential ideas of the English philosopher Herbert Spencer, who explained that collective behavior was chiefly determined by culture. The Yale sociologist William Graham Sumner and other followers of Spencer were convinced that inexorable cultural forces governed human activity, and argued that a society's institutions derived from a preexisting and static set of mores. Essentialistic cultures determined the character of law and government.[11]

Sumner's belief in enduring, fixed cultures was vividly expressed through his depiction of "oriental" and "occidental" civilization. In his massive study, *Folkways* (1906), Sumner announced that the "oriental" and the "occidental" were the "two great cultural divisions of the human race," and explained that each had complete internal consistency and "its own philosophy and spirit." Separated from "top to bottom," the oriental and occidental worlds confronted each other over an enormous gulf of "different standpoints, different ways, and different notions of what societal arrangements are advantageous." In case his emphasis on the profound and permanent divide between orientals and westerners be underestimated, Sumner imaginatively evoked, "If two planets were joined in one, their inhabitants could not differ more widely as to what things are best worth seeking, or what ways are most expedient for well living."[12]

Many nativists saw immigration in the framework of worldwide racial and cultural patterns that were unchanging and hierarchically arranged. Most importantly, this globalist vision was unmediated by an understanding of the intervening role of assimilation. Harvard historian

Henry Adams concluded, "All the so-called primitive races, and some nearer survivals" represented standing reproaches to the validity of the "most obstinate convictions of evolution." The eminent geologist Nathaniel Shaler pondered surveys of the European peasantry and decided "the inheritances of a thousand years or more" in the Old World could not be counteracted by American influences. Shaler concluded that the simple truth was that "a man is what his ancestral experience has made him." Henry Cabot Lodge, first as a U.S. representative and then as a U.S. senator from Massachusetts, served as a theoretician and political point man for the Immigration Restriction League, becoming convinced that the "lower" races and cultures resisted all efforts by the "higher" to uplift them. Lodge proclaimed that historical science demonstrated "ineradicable differences between the races of man" and ordained that the "new" immigrants sprang "from races most alien to the body of the American people."[13]

Another leader of the Immigration Restriction League, Prescott F. Hall, wondered nostalgically whether Americans wanted their country "to be peopled" by the earlier arriving "historically free, energetic, progressive" races of Britain, Germany, and Scandinavia or by the newer "Slav, Latin, and Asiatic races, historically down-trodden, atavistic, and stagnant." Hall and other nativists drew a portrait of the southern and eastern European "new immigrant" that converged with images of the "Asiatic" from China, Japan, and India. The nationalist theologian Josiah Strong claimed that Asian societies were highly organized but that "the development of the individual was early arrested, hence the stagnation of everything." Strong also described the culture of southern European immigrants as "Egyptian, Oriental, Roman Catholic, highly organized, autocratic, stagnant."[14]

Nativists who viewed the American core culture as immutably Anglo-Saxon and immigrant cultures as permanently alien logically rejected the possibility of assimilation. Jacob Gould Schurman of Columbia University, who was cited in the congressional debates over limiting immigration through discriminatory national-origin quotas, explained clearly how the new immigrants failed to assimilate and produced a balkanization of American life. They existed ominously in a deeply private world of enclaves and transnational culture:

> The public has awakened from the delusion created by the shibboleth of the "melting pot." It is disquieted and disturbed by the spectacle of immense alien communities ... more or less self-contained speaking many foreign languages, containing an influential foreign-language press, with their own banks, markets, and insurance companies and sometimes with separate schools—unleavened lumps of many European nationalities, unchanged masses of foreigners intrenched in America, yet not of it, owing in many

cases foreign allegiance, and, in general, tied to foreign countries by their language, their sympathies, their culture, their interests, and their aspirations.[15]

In the nativist imagination, the supremacist notion of Anglo-Saxon civic ownership and guardianship of American institutions dovetailed with the perception of ethno-cultural limitations besetting new immigrant groups. Stanford historian of education Ellwood P. Cubberley wrote authoritatively that "the new immigrants" from southern and eastern Europe were "illiterate, docile, often lacking in initiative, and almost wholly without the Anglo-Saxon conceptions of righteousness, liberty, law, order, public decency, and government."[16] The arguments of Clinton Stoddard Burr in *America's Race Heritage,* quoted in the House of Representatives in its climactic 1924 debate over the bill to establish discriminatory quotas against "new immigrants," exhibited how American culture was seen as a group possession of the Anglo-Saxon race to be defended against the incursions of lower races. Since "the heritage of freedom" was "the priceless heritage of a great idea conceived by the Nordic people," Burr asked, "Why do we still allow the dregs of southern and eastern European nations to swarm into our community by the thousands every day . . . ?" He warned that this unchecked flood "imminently threatens the stability, genius, and promise of achievement of the American Commonwealth . . . ."[17] Burr spoke for many nativists who envisioned representative democracy as an Anglo-American possession that could not be shared with outsiders without risking its corruption or even destruction. The 1924 report of the House Committee on Immigration and Naturalization calling for discriminatory quotas echoed this argument, concluding, "The basic strain of our population must be maintained" if the principle of individual liberty under the Constitution is to survive.[18]

Historical Perspectives and Historical Implications

The juxtaposition of public discourse on immigration in the era of national economic integration a century ago with the immigration debate in today's era of global economic integration suggests that these can be seen as two comparable historical moments, each reflecting a similar context of problematic societal conditions related to immigrant integration, and each producing similar interpretations fostering critical doubt and social pessimism about immigrant prospects. Like American society in the Progressive era, the current domestic scene of globalization is pervaded by a sense of unprecedented vulnerability to macroscale economic and demographic changes. And the national "conversation" about, and debate over, the incorporation of today's immigrants has been affected by interpretive themes recalling early twentieth-century critiques of increasing incompatibility, stagnation, and conflict. Thus, Jeremiah W. Jenks's emphasis on the declining potential of the economy to absorb immigrants resonates with today's arguments about immigration's depressive economic effects, Jacob Riis's picture of second generation regression is paralleled in themes of downward cultural assimilation in recent sociological studies, Madison Grant's fears of racial assimilation have resurfaced in alarms over the unassimilability of new ethno-racial immigrant groups, and Josiah Strong's appeals for the normative domination of Anglo-Saxon Protestantism have been echoed by calls for re-Anglicization to reinvigorate a national identity weakened by global immigration.[19]

Perhaps it is possible as well to see causal connections in these types of responses to immigration from the Progressive era to today, and to discern their underlying historical form and structure or "morphology," a term that has been applied fruitfully to social history. Indeed, several major historians have argued that the general morphology of American history consists of recurring cycles of change, leading to new stabilizing reactive or reactionary structures.[20] If they are part of such a cycle, recurrent moments of critical doubt about immigrant prospects might be viewed in historiographic terms by employing historian Herbert G. Gutman's model, posited in his well-known article, "Work, Culture, and Society," of recurring cycles of immigration-related societal change.[21] From the vantage of the twenty-first century, it may also be fruitful to go beyond Gutman's historiographic compass, which centered on a national industrial revolution in the United States, and to ask if the recurrent cycles he identified were nested within a larger and longer shift—a globalization process in which the U.S. and North Atlantic industrial revolutions constituted only an earlier phase of its unfolding.

Gutman concluded his article by summarizing, "These pages have fractured historical time, ranging forward and backward to make comparisons for several reasons. One has been to suggest how much remains to be learned about the transition of native and foreign-born American men and women to industrial society, and how that transition affected such persons and the society into which they entered." Gutman reiterated the historical and sociological centrality of this transition and its recurrent power:

> Enough has been savored in these pages to suggest the particular importance of these transitions in American social history. And their recurrence in different periods of time indicates why there has been so much discontinuity in American labor and social history. The changing composition of the working population, the continued entry into the United States of nonindustrial

people with distinctive cultures, and the changing structure of American society have combined to produce common modes of thought and patterns of behavior.

However, Gutman noted that the connections between people in transition are often not evident to themselves. People enclosed in their own era react to macrostructural shifts without consciousness of a deep historical nexus underlying societal change and how it is interpreted contemporaneously, a nexus that connects them with their contemporaries and their predecessors:

> But these have been experiences disconnected in time and shared by quite distinctive first-generation native and immigrant industrial Americans. It was not possible for the grandchildren of the Lowell mill girls to understand that their Massachusetts literary ancestors shared a great deal with their contemporaries, the peasant Slavs in the Pennsylvania steel mills and coal fields. And the grandchildren of New York City Jewish garment workers see little connection between black ghetto unrest in the 1960s and the Kosher meat riots seventy years ago.

Gutman pointed out, "A half century has passed since Robert Park and Herbert Miller published W. I. Thomas's *Old World Traits Transplanted*, a study which worried that the function of Americanization was the 'destruction of memories.'" It may be that Americanization has had an external effect as well as the internal effect to which Gutman and Thomas adverted: Americanization may also have dulled the memory of once pervasive doubts about the capacity for success of the ancestors of today's assimilated Americans of European immigrant ancestry during the peak era of the industrial revolution, when a group of public commentators feared that a sweeping and "modernizing" societal transformation that they espoused might be impeded by a retrograde population of "new" immigrants from the underdeveloped provinces of Europe.

Conclusion

Historical research has been confirming since the 1950s that the social and life outcomes of these immigrants turned out to be very different from those predicted by xenophobic restrictionists and social pessimists of the Progressive era. Studies of southern and eastern European immigrants have shown that they effectively achieved economic mobility, formed organized households, and built functional communities.[22] With intergenerational change, ethnic factors did not disappear but became part of a pluralistic fabric in which what groups had in common had a potential to become more important than what made them different. Instead of becoming an enduring underclass, immigrants paved a road to more opportunity in the middle ranges of America's changing socioeconomic structure. New fields of skilled employment developed, and immigrants and their children flocked to them. Despite the popular anxiety that industrialization depressed the working class, the occupational structure of the late nineteenth and early twentieth century changed in expansive ways under the stimulus of a new immigration. Finally, immigrants aggressively capitalized on investment opportunities in home ownership, a powerful asset to propel intergenerational economic mobility.

As immigrant aid organizations evaluate how to promote life prospects for today's young immigrants, they cannot escape being affected by the judgments of scholars and public commentators. In the current climate shaped by advocacy scholarship and tabloid journalism, organizations providing support services to immigrants should be alert to respond critically to presentist fixations that represent today's immigration in unprecedented terms. As they try to shape positively and constructively the social pathways of immigrant adaptation, they should keep in mind the fundamental historical point that social change and group change have always been parts of a broader and deeper pattern of history—contingent, unpredictable, innovative, anomalous, and latent—below the "surface world" perceived or predicted by contemporaneous agents and public spokesmen.[23] Their predecessors in the Progressive era public sphere heard immigration critics like those of today who had a limited sense of historical perspective and who had difficulty seeing short-term patterns as phases in a larger flux, whose long-term outcome would be a creative process of open-ended assimilative change.

Notes

1. Jane Addams, *Twenty Years at Hull House* (New York: New American Library, 1960; 1910), pp. 179, 181.
2. John F. Kasson, *Amusing the Million: Coney Island at the Turn of the Century* (New York: Hill and Wang, 1978), pp. 100–101.
3. Jacob A. Riis, *How the Other Half Lives* (New York: Charles Scribner's, 1890).
4. Russell Sage Foundation, *West Side Studies* (New York, 1914), Vol. I, p. 8; cited in Stephan Thernstrom, *The Other Bostonians: Poverty and Progress in the American Metropolis, 1880–1970* (Cambridge, Mass.: Harvard University Press, 1973), p. 39.
5. Ignatius Donnelly, *Caesar's Column* (Chicago: F. J. Schulte, 1890); Jack London, *The Iron Heel* (New York: Macmillan, 1908).

6. Richard Hofstadter, *The American Political Tradition* (New York: Alfred A. Knopf, 1948), pp. 236–237.
7. Joseph F. Kett, *Rites of Passage: Adolescence in America 1790 to the Present* (New York: Basic Books, 1977), p. 150.
8. Michael J. Piore, *Birds of Passage: Migrant Labor and Industrial Societies* (Cambridge, Eng.: Cambridge University Press, 1970), pp. 146–166.
9. Robert Hunter, *Poverty: Social Conscience in the Progressive Era* (New York: Macmillan, 1904), p. 61.
10. Oscar Handlin, *Immigration as a Factor in American Life* (Englewood Cliffs, N. J.: Prentice Hall, 1959), pp. 53–59.
11. Donald Fleming, "Social Darwinism," in Arthur M. Schlesinger, Jr. and Morton White, *Paths of American Thought* (Boston: Houghton Mifflin, 1963) pp. 123–130; Herbert Spencer, *The Principles of Sociology*, Vol. II (New York: D. Appleton and Co., 1899), pp. 3–14; William Graham Sumner, *Folkways: A Study of the Sociological Importance of Usages, Manners, Customs, Mores, and Morals* (Boston: Ginn and Co., 1911; 1906), pp. 53–57.
12. Sumner, *Folkways*, p. 6.
13. Henry Adams, *The Education of Henry Adams* (Boston: Houghton, Mifflin, 1918), p. 409; Barbara Miller Solomon, *Ancestors and Immigrants: A Changing New England Tradition* (Cambridge, Mass.: Harvard University Press, 1956), pp. 93, 111–116.
14. Solomon, *Ancestors and Immigrants*, pp. 111, 115; Alexander Saxton, *The Indispensable Enemy: Labor and the Anti-Chinese Movement in California* (Berkeley: University of California Press, 1971), p. 279; David R. Roediger, *The Wages of Whiteness: Race and the Making of the American Working Class* (New York: Verso, 1991), p. 179.
15. Quote from testimony by Representative John J. McSwain on April 5, 1924, Sixty-Eighth Congress, House, First Session, *Congressional Record—House*, p. 5685, Vol. LXV-Part 6 (Washington, D.C.: U.S. Government Printing Office, 1924).
16. David B. Tyack, *The One Best System: A History of American Urban Education* (Cambridge, Mass.: Harvard University Press, 1974), p. 132.
17. Quoted from Clinton Stoddard Burr, *America's Race Heritage* (National Historical Society, 1922) in testimony by Representative John J. McSwain on April 5, 1924, Sixty-Eighth Congress, House, First Session, *Congressional Record—House*, Part 6, pp. 5683–5684, Vol. LXV-Part 6 (Washington, D.C.: U.S. Government Printing Office, 1924).
18. Quoted from 1924 House Committee on Immigration and Naturalization report by Senator James Eastland on June 2, 1948, Eightieth Congress, Senate, Second Session, *Congressional Record—Senate*, Part 5, p. 6904 (Washington, D.C.: U.S. Government Printing Office, 1924).
19. George J. Borjas, *Heaven's Door : Immigration Policy and the American Economy* (Princeton, N.J.: Princeton University Press, 2001); Alejandro Portes and Ruben G. Rumbaut, *Legacies: The Story of the Immigrant Second Generation* (Berkeley: University of California Press, 2001); Samuel P. Huntington, *Who Are We? : The Challenges to America's Identity* (New York : Simon and Schuster,

2004); Peter Brimelow, *Alien Nation: Common Sense About America's Immigration Disaster* (New York: Random House, 1995); William R. Hutchison, "Strong Objections: Another Best-Selling Author Complains about Plagiarism," *History News Newsletter* (December 5, 2005).

20. John Higham, *Strangers in the Land: Patterns of American Nativism, 1860–1925* (New Brunswick: Rutgers University Press, 1955); Rowland Berthoff, *An Unsettled People: Social Order and Disorder in American History* (New York: Harper and Row, 1971); Berthoff, *Republic of the Dispossessed: The Exceptional Old-European Consensus in America* (Columbia: University of Missouri Press, 1997); Arthur M. Schlesinger, Jr., *The Cycles of American History* (Boston: Houghton Mifflin,1986).

21. Herbert G. Gutman, "Work, Culture, and Society in Industrializing America, 1815–1919," *American Historical Review* 78 (June 1973), pp. 531–588.

22. Among the numerous studies of social mobility, pertinent works are as follows: John Bodnar, *The Transplanted: A History of Immigrants in Urban America* (Bloomington: Indiana University Press, 1985); Josef J. Barton, *Peasants and Strangers: Italians, Rumanians, and Slovaks in an American City, 1890–1950* (Cambridge, Mass.: Harvard University Press, 1975); Thomas Kessner, *The Golden Door: Italian and Jewish Immigrant Mobility in New York City, 1880–1915* (New York: Oxford University Press, 1977); Stanley Lieberson, *A Piece of the Pie : Blacks and White Immigrants since 1880* (Berkeley: University of California Press, 1980).

23. Bernard Bailyn, "The Challenge of Modern Historiography." *American Historical Review* 87(1982), pp. 1–24.

Part II

Looking into the Past: Some Lessons from the "Second Wave"

Whereas the chapters of the previous part set the stage by providing a broad historical view of basic features of the immigration process, we now turn to a specific cohort of young immigrants who arrived in this country from Central Europe in the middle of the twentieth century as refugees from National Socialism. This cohort provides fertile ground for theoretical analyses, but also for practical lessons for more recent waves of young immigrants and refugees.

In the first chapter of this part, Gerhard Sonnert and Gerald Holton present the major findings of their study of this eventually highly successful group of young refugees, whom they call the "Second Wave," and they also discuss to what extent these findings may be applicable to current waves of young immigrants.

In the next chapter, Nathan Glazer uses the "Second Wave" refugees as a case to explore the dynamics and complexities that often arise within an ethnic immigrant community when a wave of newcomers arrives. Glazer argues that, despite initial animosities, many refugees quickly merged into the American Jewish community.

The chapter that follows, which is authored by "Second Waver" Clark Abt and includes lessons for young immigrants and refugees in entrepreneurial career paths, can be read, in part, as an autobiographical case study of the general social, economic, and cultural conditions under which the Second Wave group of young refugees succeeded in America.

3

Successful Young Refugees from Central Europe—Potential Lessons for Today

Gerhard Sonnert and Gerald Holton

Introduction

In the 1930s, the German government committed what amounts to the cultural, intellectual, and scientific self-decapitation of a nation by persecuting and driving away a large fraction of the luminaries of Central European *Kultur*—often to the benefit of the countries receiving these refugees. One probably has to go back several centuries, to the learned refugees who after the fall of Byzantium came to Italy and there provided a major trigger for the European Renaissance, for a historical parallel. Among the immigrants and refugees from Germany and Austria who escaped the National Socialist turmoil and came to the United States in the 1930s and 1940s were Hannah Arendt, Hans Bethe, Felix and Helene Deutsch, Albert Einstein, Erik Erikson, Kurt Gödel, Walter Gropius, Friedrich von Hayek, Paul Hindemith, Paul Lazarsfeld, Thomas Mann, Herbert Marcuse, Erwin Panofsky, Erwin Piscator, Leo Szilard, Kurt Weill, Victor Weisskopf, and Billy Wilder—to name a few. These highly talented persons arrived as adults, in many cases already recognized and acclaimed. They brought with them powerful new ideas and ways of thinking that were of tremendous benefit and often transformed their fields in their newly adopted country. It is little wonder that a large volume of scholarly research has recognized and documented their seminal achievements and contributions.[1]

These prominent individuals (members of what one might call the "First Wave" of Central European refugees reaching the United States) were, however, not the topic of our project. Instead, we chose to focus on what we termed the "Second Wave"—the close to 30,000 children and youths who were also part of the immigration movement from Central Europe to the United States in the 1930s and 1940s, and on whom little scholarly study had existed when we began our research. Here, a different, but equally extraordinary, story emerged. On early European foundations, these young refugees had to build an American structure, often despite the trauma of arriving without parents, without command of the language, without means, and with the burden of harrowing memories. In view of their inauspicious early circumstances, one could plausibly have expected that, on the whole, this group would have spiraled down into anomie and despair. But that was not at all the case.

Our research showed that, in the long and rich history of immigration to the United States, this cohort of young refugees from Central Europe eventually became enormously successful, collectively, and made remarkable contributions to their new country (among them four Nobel Prizes in science). In this chapter, we will briefly survey some of the major findings about the young refugees' socioeconomic fate, as a group, in the United States and then discuss some of the main preconditions and causes for their collective success. (For a more detailed account, see Sonnert & Holton 2006.)

Our findings of this success story of a group of young immigrants almost immediately and quite naturally raised the following question. In view of immigration being—and remaining—a key issue that will shape the future of this nation, are there any lessons for current refugee and immigration policy that could be drawn from the cohort we studied? To be sure, we are fully aware that, in many respects, the situation of the Central European young refugees at that time was very different from that of the young refugees and immigrants currently coming to the United States in large numbers. But the fact that America today is a different place from the America of the 1930s does not make the experience of those earlier refugees entirely obsolete for any practical purpose. Comparing and contrasting a largely successful refugee group with less successful groups might help tease out the circumstances that facilitate integration, and thus advance scholarly understanding of the immigration and integration process.

Our Study Cohort

The cohort under study was defined as refugees from National Socialism who were born in Germany or Austria in 1918 through 1935 and who arrived in the United States in the 1930s and 1940s. After the National Socialists under Adolf Hitler had taken power in Germany in 1933—and in 1938, through the "Anschluss" (annexation), in Austria—they implemented increasingly harsh repressions of their political opponents, cultural dissenters, and, most of all, of the Jews for whom they had no place in their new "Third Reich." This led to a large stream of refugees, most of whom were Jewish, at least according to the National Socialists' racial definition, to European countries and farther afield. The United States became a major destination for this refugee stream, both directly and indirectly. The main center of these refugees' resettlement was New York City, and more specifically Washington Heights, which was ironically dubbed the "Fourth Reich."

The principal data sources for our study were (a) 1,571 responses to detailed questionnaires we distributed among members of the target cohort (plus survey responses from appropriate comparison groups); (b) 100 lengthy, face-to-face follow-up interviews conducted with Second Wavers; and (c) several existing representative data sets, such as those of the U.S. Census and the National Jewish Population Survey.

Socioeconomic Indicators of Success

An analysis of the 1970 U.S. Census allowed us to draw a representative picture of the whole refugee group, and to compare the refugees with American-born individuals of the same age range. We examined three indicators that sociologists traditionally consider the three main sub-aspects of socioeconomic status: (1) educational attainment, (2) income, and (3) occupational prestige. The following are some key results:

First, of the American-born persons in the relevant age group, by 1970 about 15 percent of the men had attained an educational level of four years of college, or higher, and about 8 percent had gone on to advanced education beyond college. In contrast, of the former Central European men who had arrived in the United States as youths, some 48 percent had at least four years of college—more than three times their American-born comparison group. Moreover, 32 percent had received academic education beyond college—about four times more than their American-born counterparts. For the women also, the comparison was striking. Of course, there was then a very large gender gap with respect to getting higher education. By 1970, only 8 percent of American-born women in the relevant age group had at least four years of college, and less than 3 percent had gone on to higher degrees. Again, the equivalent numbers for former immigrant women from Central Europe were much higher, 19 percent and 9 percent, respectively. In sum, the magnitude of these collective educational achievements is nothing short of remarkable.

Second, the former Central Europeans eventually did very well, too, in terms of their annual income in the United States. According to the 1970 Census, the annual income of the Central Europeans who had arrived as children or adolescents between 1935 and 1944 substantially exceeded the average income of the corresponding American-born population. In fact, the average income of the former Central European men was almost twice as high as that of the American-born men of the same ages (about 192 percent in relation to the 100 percent of the American-born average). The average income of the former Central European women was also considerably elevated (149 percent), in comparison with that of the American-born women.

Third, the former German and Austrian children came also to be highly represented in the top two broad occupational categories—namely, those of professionals and managers—as defined in the U.S. Census. Again, some gender differences were evident. By 1970, two-thirds (67 percent) of the former Central European men in our cohort were in the top two categories, while 32 percent of the women from Central Europe belonged in those two top categories. By comparison, among their American-born contemporaries, only 28 percent of the men and 18 percent of the women were in the two top categories.

When we did a parallel set of analyses, in which we compared the refugees only with white native borns (to take into consideration the effect of race), the same general picture appeared, with only slightly reduced differences. In addition to comparing the immigrant cohort with the general American-born population, we also compared Jews who had immigrated from Central Europe with American-born Jews, by obtaining data from the National Jewish Population Survey of 1970. It is well known that the Jews born in America have, on average, been successful in terms of socioeconomic accomplishments (see Foner, in this volume). But even when comparing the immigrant Jews with American-born Jews, the collective educational achievements of the Central European immigrants were found to be greater, although the gap was narrower than the previously reported one between immigrants and the American-born population at large. In terms of occupation, the differences found between former Central European Jews and American-born Jews, though again, at a less pronounced scale, replicated the differences found in our earlier comparisons. In terms of income, however, no significant difference was found between the newcomers and the American-born Jews.

For a quantitative measure of top achievement, we looked into the pages of "Who's Who"; there we found about 15 times as many members of the refugee group as one would expect, given the number of refugees and the overall probability of all Americans of the same age group to be listed in that compilation.

Causes of Success

There are, of course, numerous and varied factors that all contributed to making these positive outcomes possible. One way of ordering them might be to distinguish four types of factors: those of a general historical nature that affected the American population as a whole, those pertaining to the refugees as a group, those on which there was internal variability *within* the refugee cohort, and finally those related to the efforts of aid agencies. Here, in keeping with the theme of this volume, we consider them under the aspect of applicability to the current situation.

Advantageous Framework Factors

When the first refugees started to arrive in the United States in 1933, America was still mired in the Great Depression. However, the economy recovered and boomed in the war and postwar years. Many of the young refugees began their careers under these generally favorable circumstances (as did the American borns of the same age cohort). The state of the national economy has been—and remains—an obvious and powerful determinant for both the size of the immigration stream and the economic opportunities open to young newcomers.

While there was an economic boom in general, the academic and scientific employment sector experienced a particularly large expansion. Thus, there was a match between the sociocultural image of a desirable job that had been imparted to the young immigrants and the actual availability of such jobs. Our data showed that a relatively large proportion of the former refugees found employment in this sector. The weakening of anti-Semitism, especially in this area, also helped, as did the growing relevance of the broad, generic category of "white Americans."

Whereas the current economic situation and the current American occupational structure are different, specific smaller immigration streams now of highly skilled newcomers and their young family members might experience a similar fit as well as positive socioeconomic outcomes (e.g., perhaps in computer and IT fields).

The fact that the American educational system of the day enabled the young refugees, as a group, to achieve educational goals at the highest levels played a pivotal role in the successful integration of these young refugees. Especially, functioning and high-quality institutions of public education in New York City (where a large number of the refugees had arrived) had a substantial beneficial impact. The young refugees' access to these institutions tremendously helped their settling in and succeeding in their new country.

Just as it had exerted a potent effect for our cohort, the educational system remains a key determinant for young refugees' outcomes today, a determinant that cannot be overestimated. Hence, this is a prime strategic target for study (see the chapters by Suárez-Orozco & Gaytán and by Perlmann, in this volume) as well as for intervention, not only for the school systems themselves, but also for state and federal government and NGOs. In the 1930s and 1940s, the public educational system in New York City, but also elsewhere, was demonstrably able to deliver positive outcomes to these young immigrants. It did so not necessarily by making special allowances for them, but by providing the same opportunity to a solid education as to the other youths, on which the young refugees then were able to capitalize to a particularly high degree (partly owing to a good educational background from Europe). Education researchers, practitioners, and policy makers might well profit from looking into this educational success story more closely.

Factors Pertaining to the Refugees as a Group

For many immigration streams in the past, assimilation was the prime goal (being a normative expectation held by American society as well as by most immigrants themselves). The young refugees of our cohort conformed to that expectation and achieved assimilation to a particularly high degree. The horrendous circumstances of their expulsion from Central Europe weakened "old country" ties more perhaps than among other groups of young refugees or immigrants. Most of our participants were not "exiles" with the dream of returning to their homeland. For them, it was clear from the very beginning that America was their new home, where they had to sink or swim. Typically, they immediately committed to making a living in this country. This immediate and unequivocal commitment, driven by the perceived lack of any alternative, is likely to have contributed to their eventual success.

This general theme of assimilation can be illustrated through the example of language acquisition, perhaps its key dimension. The young refugees put great efforts into learning English as quickly as possible, and they, on average, succeeded speedily and with ease in that endeavor. Our group is interesting in that it represents perhaps an extreme case of an immigrant group's willingness to adopt the English language. Linguistic assimilation was encouraged and indeed expected by government agencies as well as civil society—this applied to the Central Europeans just as to other immigrants of that era. To this, one special factor was added: German was the language of an enemy in World War II, so that being identified as German was certainly not desirable in the early years of the refugees' settlement.

There is now a greater tolerance for, and even normative celebration of, diversity of origin. Nonetheless, even today, quick language acquisition may be a key to young immigrants' success in the larger society (see Suárez-Orozco & Gaytán, in this volume).

Assimilation, however, was only part of the Second Wavers' story. We found the concept of cultural capital (Bourdieu 1983) particularly powerful in helping explain the young refugees' success. Two types of cultural capital should be distinguished here: cultural capital that applies diffusely to almost everybody in a certain national or ethnic group, and cultural capital that is specifically tied to the higher social classes. As to the first type, almost all the former refugees (except those very few who arrived at a very young age and without family) were culturally competent in German and Central European customs and language. They could, and many did, use this cultural capital for making a career in this country by working as some type of cultural intermediary—as a language teacher or in the military for the occupation forces or in business by being involved in European operations or markets. This may be true, to some degree or other, for all immigrants. But our immigrants were in the favorable position that, in several important aspects, the Central European currency of cultural capital was "convertible," that is, recognized also in America. The American academic, intellectual, artistic, and scientific fields were to some extent modeled after the Central European core concept of "Bildung" (roughly meaning the process and outcome of a liberal arts education)—hence quite familiar to the young refugees. Moreover, whereas in many cases the culture of immigrants tends to be devalued in America as premodern, irrelevant, or even inferior, there existed at that time varying degrees of respect for the Central European "Kultur" in America. (A substantial fraction of the American cultural elite had been to Central Europe for part of their training.) Denouncements of cultural inferiority that critics of immigration habitually leveled against newcomers simply lacked plausibility in this case. Moreover, coming from highly developed and industrialized home countries and, in many cases, being accustomed to life in big cities may have also reduced the magnitude of the "culture shock" experienced by the young immigrants.

The situation of most current young immigrants, in terms of cultural capital, is certainly different from that of the Central Europeans. But it may be very advantageous for those who help young immigrants succeed to look carefully into what cultural capital the young refugees bring with them, and if and how it might be converted into American currency. We should also keep in mind that even though the young refugees of the "Second Wave" assimilated to a great degree, they retained some elements of their European upbringing in terms of cultural capital. These elements were, on the whole, not a handicap but a *distinctiveness advantage* that helped them succeed in America. In the current climate of greater acceptance of cultural diversity, it might be both feasible and profitable for young immigrants to capitalize on their distinctiveness advantage (see Waters and Halter & Johnson, in this volume).

Another interesting, but perhaps not surprising, fact that emerged from our study was that many refugees who attended American institutions of higher education found initial intellectual sponsors from among members of the older group of immigrants, that is, the First Wave. Sociologists of immigration have long focused on the ethnic niches in which newly arrived immigrants cluster, and which make available crucial *social capital* for their settling into the new country. The prevailing opinion among those scholars ascribed two characteristic features to the ethnic niches: they were considered a temporary phenomenon that would fade away as immigrants became more established, and they were thought to be populated predominantly by immigrants with no or few skills. Newer research, however, found that those niches are more permanent in nature, and that they are used also by relatively high-skilled immigrants (Waldinger & Der-Martirosian 2001). In our case, the academic system—through the nexus between First and Second Wavers—constituted some form of ethnic niche at the high end of the skill spectrum.

While most of our study participants reported that they had never considered a life of deviance or crime, a few said they came close to it in their younger years, but were somehow pulled back from the edge through resolute family interventions. In other words, among those refugees who arrived with their family, the family as an agency of authority and socialization was typically strong enough to suppress any peer-related tendencies toward deviance.

To be sure, the refugee families were not devoid of a dynamic often found among immigrants. As one of our participants' stated matter-of-factly, after arrival "I was the parent and they were the children." Typically, this shift in family dynamics and the resulting potential for generational conflict were resolved by the young immigrants' increased efforts to take care of, and even provide for, the family, rather than by a split from the family. Those concerned with helping young immigrants succeed are well advised to find ways to strengthen family structures or address their absence with suitable substitutes.

Our participants tended not to think of themselves as victims or to portray themselves as such. It was patently obvious that among all the victims of the Holocaust, their fate was still relatively benign (e.g., as compared with those who were murdered or with the Concentration Camp survivors). They, thus, as a group possessed a high sense of responsibility for their own life and a strong determination to make it on their own, without the expectation of special consideration.

Complaints about suffered injustices and hardships are now received more sympathetically in segments of society and the mass media. This change, although it sounds beneficial, may actually undermine the psychological drive toward self-reliance that certainly helped our young immigrants succeed. The challenge for aid organizations here is to help in a way that starts young immigrants on a productive career trajectory in this country, but does not undermine their initiative.

The young refugees typically had a great psychological drive to succeed. On the whole, the life stories of the young refugees in our study vividly illustrated the topos of the "hungry immigrant" who is highly motivated and determined to succeed in the new country despite all obstacles and disadvantages (see Waters, in this volume). But there was often a cost to their success, a cost that caught up, sometimes only in later life. In the course of our research, we also learned of psychological discomforts, suicides, and other tragedies that might be attributed to early psychological traumatization.

From a larger perspective, thus, the outcomes for this group of young refugees can be described with the motto "privatized costs, socialized benefits." As a group, the young refugees went on to make solid contributions to this country, so that society has benefited greatly from them. On the other hand, the psychological cost that has also been involved for many refugees has been borne quietly and privately.

Factors with Variability

A particularly important characteristic of this group of young refugees is that compared with other refugee groups, exceptionally many of its members came from an elevated socioeconomic background. In the broad categories of the U.S. Census, 30 percent of the Second Wavers we surveyed had fathers who had been professionals in Europe. Thirty-eight percent of the fathers had been managers in Europe, and only 32 percent had worked in other occupations. We do not doubt that when it comes to explaining the overall success of this particular cohort of young refugees, these social background factors played a considerable role. Nonetheless, contrary to some simplifying stereotypes about the Second Wave, these data also show that *not all* young refugees came from the well-to-do Central European bourgeoisie. The variability in their socioeconomic backgrounds enabled us to control, to some extent, for the effect of those backgrounds. When we looked at only those refugees who did *not* come from elevated socioeconomic backgrounds and compared their socioeconomic achievements with those of all native-born Americans of the same age cohort, we found that even those refugees did substantially better than the native borns in all three sub-aspects of socioeconomic status. Furthermore, we were able to investigate, *within the refugee cohort,* the effect of the young refugees' class origins on their later outcomes.

The group we were studying is a strategic research site for examining the extent to which an elevated socioeconomic status can be transmitted from one generation to the next, even though the material substrate of this class position had been lost—that is, even though the parents had in some cases perished, or frequently experienced severe downward mobility in the wake of their migration to the United States, and even though most of the family wealth had typically vanished. Whereas, usually, higher-status parents provide their offspring a host of material advantages, along with values, norms, and ambitions, to help them maximize their chances in life, here we encounter the isolated effect of nonmaterial factors.

Even in the absence of the material substrate of social class, we found an enormous impact of their family's social class in Central Europe on the young refugees' occupational outcomes, despite the status discontinuities associated with flight and resettlement. This is probably one of the most important results of the project. The father's occupational status in Europe predicted the young immigrant's own eventual occupational status, and there was even an *additional* effect of other forebears' occupational status. Thus, our study demonstrated the strong effects of the socialization of a class role, or of cultural capital of the second type—cultural capital tied to social class. This may have repercussions for the work of aid organizations, which will be discussed later.

Regarding their age at arrival, we found among our cohort of young refugees that younger arrivals did better than older arrivals in terms of educational as well as occupational outcomes. These statistics were somewhat surprising because numerous participants who were old enough to have attended the *Gymnasium* (secondary school) mentioned that the superior academic training they had received there allowed them to thrive in American institutions of education. However, the statistical approach revealed that even with this advantage, they did not do as well, collectively, as the younger ones did. This finding—which applied to young immigrants with a particularly strong academic background, on average—suggests much more dramatic disadvantages for school-age immigrants who come with weak or nonexisting educational credentials. Hence, particularly strong needs for aid and support may exist among young immigrants of school age or older.

Factors Pertaining to Aid Organizations

Our study showed the expansive and largely decentralized network of nongovernmental immigrant aid organizations to which the refugees could turn. Altogether, according to Julius and Edith Hirsch (1961: 51), as many as 110 different nongovernmental organizations, Jewish and non-Jewish, were involved in helping the refugees, including children—a remarkable proof of the typically American tendency for private citizens' action. A few of these organizations, especially the bigger ones, were preexisting. Some, such as HIAS (Hebrew Immigrant Aid Society), had originated to help an earlier wave of Jewish immigrants from Eastern Europe who fled persecution in the late nineteenth and early twentieth centuries and now (as that immigration stream had all but dried up) stood ready to spring into action for a new group of arrivals that, in a sense, provided those organizations with a continued *raison d'être*. Among other organizations with a long and international history of aid to refugees is the Society of Friends (Quakers), remembered perhaps best by its having organized "Kindertransporte" from Central Europe to Britain.

In an important sense, the success story in the United States of our cohort of young refugees is equally the success story of nongovernmental refugee and immigrant aid organizations, such as those represented in this volume, of which they can be justly proud. We are happy to communicate a tremendous feeling of gratitude still felt and expressed by many of our study participants for the help, financial or otherwise, that they received over half a century ago from all the various aid sources. The possibility to access this extensive and complex network of support was a key factor in the eventual success of this wave of youth.

A notable difference in this respect between the 1930s and 1940s and today is the increased involvement of the government. Many advocates for refugees and immigrants today lament cuts in refugee funding by the government and criticize what they consider an insufficient level of government involvement in the issue. From a historical perspective, however, even the relatively limited involvement of the government today exceeds the government aid available to the Second Wavers. To the extent that the government plays a more active role in helping young refugees and immigrants, the role of the NGOs changes., It now becomes ever more important to know about, access, and steer clients to, the various government programs.

Recall that one of the most powerful predictors of eventual socioeconomic success *within* our group of young refugees was the father's—and, in addition, the grandfather's!—social class in Europe. No aid organization or government can provide young refugees with a family tree of choice, but still it might be important for aid organizations to be aware of the forceful mechanisms of intergenerational status transfer and of the power of cultural capital. Young immigrants and refugees with certain class roles and a high amount of class-based cultural capital have a relatively high probability of success within the opportunity structure of American society. They are also the most likely to look for help, and to leverage support to their own advantage. This may lead to a *paradox of aid efficiency*: an aid institution that preferentially supports those who are likely to be rather successful anyway can assemble the most impressive institutional success record. The institution might make more of a difference, but might look worse on the surface, if it focused on those who lack cultural capital.

Conclusion

In our study we found that as a group, the young refugees who came from Central Europe to the United States in the 1930s and 1940s eventually had very successful life and career paths in their new country. The story of the Second Wave is an account of cataclysm, traumata, resolve, adaptation, resilience, and, on the whole, extraordinarily productive lives, and we do believe that something of relevance can be learned from our study to increase the chance for successful lives for young refugees and immigrants who today arrive in this country.

While there were several general factors that helped this positive outcome, there were also some that specifically benefited this group of young immigrants, among them their rapid language acquisition, cultural and social capital, psychological resilience, and access to a vast network of nongovernmental aid organizations. Within the cohort of refugees, *cultural capital* tied to the family's previous socioeconomic status in Central Europe played an important role in the young refugees' eventual occupational outcome. *Young age at immigration* was also associated with positive outcomes.

On the basis of these findings, targets for intervention that might benefit current young refugees and immigrants are, primarily, the areas of education, of language acquisition, and of strengthening family structures. Furthermore, there should be a particular focus on those of lower-class backgrounds, who may not be able to avail themselves of opportunities and of assistance as easily as might young refugees and immigrants from more elevated class backgrounds.

Note

1. For example, Ash & Söllner (1996), Coser (1984), Fermi (1971), Fleming & Bailyn (1969), Heilbut (1983), Hughes (1975), Ingrisch (2004), Jackman & Borden (1983), Kent (1953), Möller (1984), Sokal (1984), Spaulding (1968), Stadler (1987, 1988), and Strauss, Fischer, Hoffmann & Söllner (1991).

4

The Second Wave and the American-Jewish Community

Nathan Glazer

The research enterprise that has occasioned this volume has described itself, in the course of the years during which it was conducted, as a study of the "Second Wave" (Sonnert & Holton 2006). By that it meant the young Jewish refugees who came to the United States from Central Europe in the 1930s and early 1940s, as distinguished from the adult and mature refugees concerning whom there has been a great deal of research. The experience of this generation raises many interesting questions. The issue I wish to address is, how did this group of young refugees eventually adapt to, relate to, and perhaps influence the Jewish community that was already established in the United States?

Waves of Jewish Immigration to North America

Those of us who study the American-Jewish community are well accustomed to the terminology of "waves of migration" in characterizing its history and development. American-Jewish history is generally considered as being formed by three distinct and distinctive migrations or waves of settlers, migrants, immigrants—the terms we use change over time, and with the character and motives of those who have come. The first wave—scarcely a wave in size, but important in establishing the first communities of Jews in the colonies that were later to become the United States—comprised the Spanish and Portuguese Jews who came in the seventeenth and eighteenth centuries, variously from Brazil, the Netherlands, England, the West Indies, and other places to which they had scattered.

They established the first Jewish communities and synagogues in the Dutch and English colonies of the Atlantic coast.

The second wave was the German-Jewish migration of the nineteenth century, which spread throughout the broad expanse of the expanding United States, established Jewish communities everywhere, and introduced a revolution into the Jewish religion, bringing—and expanding here in the United States—Reform Judaism. The third wave was the East European migration of the period from the 1880s to the 1920s, brought to an end by the restrictive immigration legislation of the 1920s.

We do not speak of fourth waves or a fourth migration, even though there has been subsequent Jewish immigration to the United States, because Jewish immigration since the 1920s has been, in scale or influence, relatively minor compared with the first three. Since the great East European wave was brought to an end in the 1920s, we have had first of all the German-Jewish refugee immigration from 1933 to the mid-forties, numbering at most 150,000, including the children born abroad studied by Sonnert and Holton. There was a second wave of survivors of the Holocaust, predominantly East Europeans who came in the late 1940s and 1950s. Then there have been a substantial secondary immigration from Israel itself, a migration of Iranian Jews after the revolution of 1976 there, and a rather substantial immigration of Russian Jews since the weakening of Communist Soviet restrictions on emigration in the 1980s and in particular since the collapse of the Soviet Union (see Kasinitz and Gold, both in this volume).

But all these, compared with the huge East European immigration of roughly two million persons in the period from the 1880s to the 1920s, are better to be considered wavelets rather than waves. They have had only a modest influence on the American-Jewish community itself. They arrived to find a community already of such size and elaborate institutional development that their impact could only have been marginal. Nonetheless, of all these immigrations of the last 75 years, the German-Jewish refugee migration has been, I believe, the most important and the most significant in its impact on the American-Jewish community. It was a unique migration because of its distinctive relationship as a refugee community to the established American-Jewish community.

Immigrants in general, and refugees in particular, commonly are persons who have to overcome difficulties—of language, of adaptation to a new culture and society, of inferior or perhaps no education, and with skills inappropriate to making a good life in a different country. It is hard for them to make use of talents and training that are of little or no value in a new country. They have been raised with a different education—or in the case of many current immigrants and refugees, hardly any education at all.

Stereotypical Views and Their Shortcomings

Much of this applies to German-Jewish refugees, too. But what was unique in the wave of adult refugees in the 1930s and 1940s was that in education, in achievement in professional roles of high status, the German-Jewish refugees were superior to the American-Jewish community of the 1930s. Whereas each subsequent wave of immigrants past the first saw itself as inferior to the one that preceded it, in the case of the German-Jewish refugees of the 1930s, this relationship was reversed. They saw themselves, and were seen by the American-Jewish community of the time, as superior in education and social status.

A consciousness of this superiority and of complementary inferiority was in fact steadily wedged in the minds of both the adult German-Jewish refugees and their hosts in the American-Jewish community. It could not but play a significant role in their adaptation to American and American-Jewish life.

The American Jews of the 1930s and the German Jews of the 1930s were at very different stages in their adaptation to the societies in which they lived. The American Jews of the 1930s were originally overwhelmingly East European, and most were of the working class. They were rising educationally and occupationally, but most were hard hit by the Depression, as all Americans were. It had been only a decade or so since the East European Jews had begun to display a striking capacity to seize the opportunities available for higher education, to a degree that no other working-class community could or did (see Foner, in this volume). As one result, as is well known, in the 1920s many American institutions of higher education and professional education instituted "Jewish quotas," which prevailed in the 1930s when the German-Jewish refugees were arriving.

There was a smaller German-Jewish stratum in the United States, earlier migrants from Germany of the nineteenth century, who had established themselves in retail trade, reaching up to the ownership of major department stores, and in the professions, including some leading bankers. They already had substantial influence in American politics and in parts of the American economy. But within the Jewish community, they were a small minority, swamped by the much greater number of East European Jews, and in conflict on various issues with them.

The German-Jewish refugees had some connection through family relationships with the older, established German Jews of the United States, but these connections had frayed over the long period since the end of major German-Jewish emigration. Many of these distant relatives helped in providing guarantees that the new refugees would not become public charges, a key step in securing the scarce visa necessary for immigration. Indeed, one reason German Jews seemed superior in social status was that those who were able to emigrate were often indeed the better off, the better educated, the more prosperous. The poorer German Jews tended to be left behind to face the horrors of the Holocaust. Had more been able to get to the United States, they might have found more in common with the American-Jewish community of the time.

But once arrived in the United States, German-Jewish refugees had to depend for help on the established or newly created Jewish institutions, or were on their own. The American Jews they met, and the employees and leadership of the institutions with which they dealt, were for the most part East European Jews of the last great migration rather than the descendants of the German Jews of the earlier migration.

How did the perceptions of the two groups toward each other affect the adaptation of German-Jewish refugees and their children? Jack Wertheimer, a child of German-Jewish refugees who became a professor of American-Jewish history at the Jewish Theological Seminary, has vividly pictured the common American-Jewish view of German Jews:

> For the preponderant majority of American Jews who gave the matter any thought at all, the history of German Jewry serves as a powerful cautionary tale. Some time before the Holocaust, according to this folk wisdom, there lived in Germany a Jewish population that was more assimilated than any other in the world. The Jews of Germany distorted or hid their Jewishness in a desperate effort to win the acceptance of their gentile neighbors. They stifled all feelings of *ahavat yisrael*, love of fellow Jews, and instead treated their coreligionists, particularly East European Jews, with contempt and ridicule. And in their bearing, dress and cultural outlook they were "more German than the Germans." ... German Jews assimilated in an unprecedented manner. And then they were punished brutally.... Jews were taught a lesson that all Jews must remember: assimilation cannot work; the only protection Jews have is to concern themselves with the fate of their coreligionists. The experience of German Jews teaches us all that escape from Jewishness is impossible.[1]

As an American Jew of East European Jewish immigrant parents, I can vouch for the accuracy of this picture of German Jews among East European Jews, and must shamefacedly admit that it was not only among the ordinary but also among the best educated that one could find this picture, or caricature, to be taken as simple truth.

In fact, as in all caricatures, there was some connection to reality. A much higher percentage of German Jews than American Jews intermarried in the years before Hitler. In contrast, intermarriage was rare for

American Jews in the 1930s. Many German Jews were university graduates, whereas this was rare in the 1930s among East European Jews. Many German Jews converted, while conversion was also minuscule among American Jews. German Jews, despite the presence of anti-Semitism, could become distinguished professors in German universities before 1933. In contrast, this was a time when Jewish professors in American universities were a rarity. Thus, German Jews were indeed more assimilated, if we use some conventional measures of assimilation.

But it is true that we ignore the complexity and variety of German Jewry in noting all this. As Wertheimer continues, after giving this accurate caricature of the American-Jewish view of German Jews:

> It does not give me any satisfaction to relate this cautionary tale, for I view the German-Jewish experience in a markedly different manner. As the offspring of religiously observant, Jewishly active and affiliated refugees from Nazi Germany, I hardly regard German Jewry as the assimilated Jewry *par excellence*.[2]

The caricature of German Jews in the East European Jewish mind must be modified in two respects to take account of a much wider range of German-Jewish experience in Germany than is normally recognized. First, the highly educated professionals and academics did not constitute all of German Jewry. One must not forget the Jewish communities in the small market towns who preserved a simple Orthodoxy, the many small shopkeepers of the cities, and the artisans of various kinds. Second, even within the highly educated stratum, outlooks were diverse. As against the American-Jewish caricature, many of the intellectuals and artists of the flourishing Weimar (and Viennese) avant-garde culture maintained some knowledge of and connection with Jewishness and perhaps Judaism, some indeed much more than this. Martin Buber, Franz Rosenzweig, Gershom Scholem, and many others devoted their intensive German educations and their talents to explorations in Jewish history, religion, and mysticism. Our view of Weimar Germany should include not only the host of writers and artists and scholars who have shaped modernism and had no connection aside from genealogy to Jews and Jewishness, but also the exemplary figures who have done so much to connect Jewish traditions, religious and secular, to modern thought.

Consider the close friendship between Walter Benjamin and Gershom Scholem, now explored in a number of volumes. Both were remarkable products of German-Jewish education and upbringing, but one of these men we today enshrine as the very emblem of the Weimar intellectual, while the second moved to Palestine to become the greatest scholar of

Jewish mysticism and Kabbala. Scholem steadily urged his friend to join him there, but Benjamin to his misfortune did not. Benjamin's other dear friend, Theodor Adorno, connected him to the Frankfurt group of social scientists and philosophers headed by Max Horkheimer, and their Institut für Sozialforschung, with its endowment from an Argentine Jewish millionaire, could sustain Benjamin in Paris as a refugee writer. Benjamin remains a fascinating bridge figure for German Jewry. While Adorno helped maintain his grants from the Frankfurt group in exile, Scholem made arrangements for him to emigrate to Palestine. Another close friend, Hannah Arendt—who was in time to edit the first collection of Benjamin's writings to appear in English and thus launch his remarkable posthumous career as a prophet of postmodern culture—was also in Paris, sustained by her work for a Zionist organization, and deeply involved in the politics of Zionism.

Benjamin's story—and the stories of many others—tells us that the divorce we make between the larger Central European culture, whether of Berlin or Vienna, which so many of us admire and recognize as a principal shaper of modernism, and Jewishness, cannot be so easily maintained.

Jewish life, despite the pace of assimilation, was alive and strong in Germany, and while the fact is not much noted, it played a great role in affecting Jewish life in the United States. Despite the overall picture of assimilation, a distinctive German-Jewish identity existed, and German Jews were creative in Jewish religion and Jewish culture as well as in the larger German culture with which they so fully identified. It was in Germany that the effort to connect the old Jewish religion, hardly changed for a millennium, to contemporary life was attempted: Orthodoxy was modified, and whereas one stream promoted an Orthodoxy related to modern life, another explored a sharp break with traditional Judaism while still maintaining the Jewish religion. Both were carried to America, and did much to reinvigorate both streams here.

Even those segments of the community that were well along the path of assimilation and of reducing their Jewish heritage to irrelevance were forced by a few years of Nazi brutality to engage more with their Jewishness. Jewish children were excluded from German schools, and many went into Jewish schools, some newly founded by Jewish teachers who had been expelled from the German schools. The persecution to which these children were subjected often strengthened their interest in their Jewish background.

German-Jewish refugees came with a brief but strong and harsh education in what it meant to be Jewish, and many seized on one or another aspect of their heritage. The connections between the great German-Jewish intellectuals we admire and their Jewishness are not much explored, but whenever we look at these figures—Albert Einstein and a host of scientists in many fields; political philosophers like Hannah Arendt, Leo Strauss, and many others; psychologists and psychoanalysts like Kurt Lewin, Erich Fromm, and many others—we find that their connection with Jewishness is rather more than genealogical or biological. The assimilation story we tend to impose in our studies of American ethnic groups is rather more complicated than a simple progression from an ethnic identity to some general "American" identity. And so it was with German-Jewish assimilation: there was much in it that remained distinctively Jewish, and these Jewish aspects emerged more sharply under the Nazi regime and after the migration to the United States.

Integration

One might have expected, looking forward from the 1930s and 1940s, that the odd consciousness of superiority and inferiority that characterized the relationship between the German-Jewish refugees and the greater part of the American-Jewish community would lead to a separation between the refugee community and the settled American-Jewish community. After all, one notes a substantial degree of separation today between the American-Jewish community in general and the other wavelets of the last 50 years—most markedly, the Hasidim, but also the Israelis, the Iranians, and the Russians.

Language, culture, and degree of religiosity, all divide them in various degrees from the generality of American Jews. And if the American-Jewish view of the extreme assimilation of German Jews into German culture and society was accurate, a separate development for German Jews should have been reasonably expected.

Yet I believe the most important fact about the Second Wave was that this separate development did not happen: the children and adolescents who had come from Germany were eventually fully merged into the American-Jewish community, often indeed in leadership positions. They had become indistinguishable within the larger American-Jewish community. The distinctive institutions the German-Jewish refugee community had created have for the most part disappeared, their neighborhoods—primarily Washington Heights in New York—are gone, and they have become American Jews, with only modest remnants of their difference surviving. That is what the Sonnert-Holton research on the Second Wave tells us. There are many details to be teased out in subsequent research; the data, gathered with skill and ingenuity, are all there. But note a few findings that suggest that the German Jews of the 1.5 generation have become very much like other American Jews.

Both the young refugees and their American-born Jewish counterparts have done well in terms of education, occupation, and income. If the Second Wave has done somewhat better than the American Jews among whom they arrived, it can be most directly explained by the fact that their families' educational and occupational background was on the whole of a higher level—there were far fewer workers among their parents than was the case in the American-Jewish community at the time, and many more educated professionals.

But let us consider some of the other findings, for instance, those on intermarriage with non-Jews. Intermarriage was certainly considerably higher in Germany in the 1920s and early 1930s (pre-Hitler) than among American Jews at the time. But it was scarce among the Second Wave— only 12 percent chose non-Jewish spouses, certainly higher than for their American co-religionists at the time, but not very high (almost half chose spouses who were also refugees). I noted that 62 percent are members of synagogues or temples, about the same percentage that the study found among the American-born Jewish sample with which the second-wave sample was compared. Surprisingly, despite the common identification of German Jews with Reform, a higher percentage are members of Orthodox synagogues—16 percent as against 8 percent. And they claim to attend more religious services a year than the comparable American-born sample—20 percent as against 14 percent. Understandably, fewer know some Yiddish than among the American born, who are of course almost all of East European background—37 percent as against 54 percent. But more know some Hebrew—62 percent as against 35 percent.

Their political attitudes mirror the liberal attitudes of the American-Jewish community. About 20 years ago, Abraham J. Peck asked about 40 German-Jewish refugees, including some of their children, to consider the German-Jewish legacy in America.[3] We learn from this that while some of the German-Jewish refugees and their children have eventually resettled in Israel—and write their reflections from there—others share the left-wing critical views about Israel that are common among liberal Jews.

Peck's enterprise evoked an interesting series of autobiographical reflections, which tell us a great deal about the German-Jewish refugees and their experiences and the changes they have undergone in America. While most of the respondents in Peck's study are professors in institutions of all kinds, some of them are in the two major institutions that educate rabbis, the Jewish Theological Seminary and the Hebrew Union College— Jewish Institute of Religion. Some of them are proud of what German Jews have accomplished in America, as individuals and in terms of building Jewish institutions. A few of the respondents pointed out that the presidents of both of the major institutions that educate rabbis were, at the time, German Jews, as was the case with the heads of major Jewish congregational organizations and other key leaders of the American-Jewish community. Jacob Neusner, the great Jewish scholar, has pointed out (in conversation) that German-Jewish refugee scholars brought a level of sophistication in Jewish scholarship to the American-Jewish seminaries that had simply not existed before.

Peck subtitled his book *From Bildung to the Bill of Rights*. He was looking for a distinctive orientation that might have survived among German Jews and their children. There was still Bildung of a sort among German Jews and their children, but Goethe, Schiller, and Heine had been replaced by the *Nation,* the *New Republic,* and the *New York Review of Books. Bildung* had been transformed in America.

Have the German Jews assimilated? Of course—but to what? More to the American-Jewish community than to any alternative. The road to assimilation in the United States is via the ethnic group, either an old one to which a new group connects, or one created anew out of the immigrant stream. The effect of the German-Jewish refugee migration in the American-Jewish community was to strengthen it and broaden it, rather than to propel it more strongly on the road to an assimilation that negates distinctive identity.

Notes

1. Jack Wertheimer, "The German-Jewish Experience: Toward a Useable Past" (pp. 233–239) in Abraham J. Peck, ed., *The German-Jewish Legacy in America, 1938–1988* (Detroit, MI: Wayne State University Press, 1989) p. 235. See, too, in Christof Mauch and Joseph Salmons, eds., *German-Jewish Identities in America* (Madison, WI: Max Kade Institute for German-American Studies, 2003), essays by Henry Feingold and Manfred Kirchheimer.
2. Ibid.
3. See Peck, note 1 above.

5

Helping Young Immigrants/Refugees Become Entrepreneurs

Clark Claus Abt

For its economic, intellectual, social, and cultural growth and progress, America has always depended, and depends now and into the future, on the rise of creative and hardworking young and middle-aged persons, including refugees—some from humble beginnings, and some from their previously persecuted or otherwise threatened status. A major propellant of this rise is successful entrepreneurship. *Webster's Dictionary* defines an entrepreneur as a person who organizes, manages, and assumes the risk of a business or an enterprise. Among these entrepreneurs are immigrants who, like their native-born peers, are company founders, daring dreamers, and creative problem solvers with confidence and energy.

Thus, in MIT's 1989 "Salute to Founders," President Paul Gray paid tribute to "the energy, confidence and vision of 99 members of the MIT family who had the courage to take risks, the knowledge to solve problems, and the creativity to dream dreams. They represent an even larger group... who collectively have established in the Boston area 600 firms, generate annual revenues of $37 billion and provide employment to over 180,000."

Of the 99 enterprise founders listed, six were young refugees or immigrants—myself among those six. A company founder several times over, I write as a representative of that fortunate subgroup of refugee or immigrant entrepreneurs who found America *the* land of opportunity for the realization of their hopes and aspirations, as well as a refuge from oppression. I have been an entrepreneur since I came to America 73 years ago and tried to start a shoeshine business at age seven, but it took me another 14 years to become self-supporting, and 25 years to become relatively successful at age 32 as a systems engineering department manager in a major aerospace electronics corporation. I followed many different and some divergent paths to later success, and I was finally able to weave together my diverse interests and skills into a thriving enterprise that created over 1,000 jobs and several technological and social innovations by the time I was 40 and—on paper—a (small) millionaire.

In a book that mainly looks at the experience of large samples, perhaps a chapter on an individual's life and his encounters with America's opportunity structures will be of interest.

The Entrepreneurial Potential of Refugees

Most important today for our goal of helping the increasing tide of refugees from war-torn areas become successful and self-supporting by obtaining marketable skills, productive jobs, and family-supporting incomes—often by starting sustainable small businesses that grow jobs and incomes and add to quality of life—are the lessons learned by persons much like me among refugee entrepreneurs. It is understood that only a fraction of the general population—perhaps 10 percent—is entrepreneurial. Historical evidence suggests that this fraction of entrepreneurs, whatever its actual size, is significantly greater among immigrants and refugees. Refugees and immigrants tend to self-select for courage, ambition, determination, energy, imagination, and initiative—or they would be content to remain in their native lands or, if persecuted, in some more nearby intermediate country of refuge. Some of the most entrepreneurially successful refugees have the greatest potential for contributing to their adopted nation's economic, social, intellectual, and cultural growth, through jobs and income creation and scientific and artistic achievements.

Therefore it makes sense to assist young refugees and immigrants to become successful entrepreneurs if they have the relevant characteristics, as many of them—a larger fraction of them than of the general population—do have. It may even be argued that it is the continual input of entrepreneurially driven refugees and immigrants (many of whom might be considered economic refugees from poverty) that has helped significantly to drive the American economy to become the most productive in the world. And this is why current restrictive American refugee asylum and immigration policies threaten America's continued superior productivity.

Why do poor refugees and immigrants, especially those without good English language skills, tend to start their own businesses? If, because of prejudice, lack of familiarity with the new country, and lack of language skills, refugees or immigrants cannot get a job, they tend to try to create their own. Also, if they cannot get a good job matching their previously acquired skills, they might venture to start their own business on the basis of their skill set. This is entrepreneurship by necessity as well as by preference.

Refugees rebounding from oppression and tyranny in their home country gain an additional thrust to grow and thrive entrepreneurially when they enter the American atmosphere of freedom. Refugees from oppression and existential threats of war thus have additional motivation to succeed in their country of refuge.

This chapter is concerned primarily with refugees from persecution, violence, natural disasters, and war, rather than with the larger population of immigrants motivated by a desire to improve their economic prospects. Of the over 12 million undocumented aliens in the United States today, probably only a small percentage would claim to be refugees—perhaps one or two hundred thousand. But worldwide, the UN has estimated that there are now some 25 million refugees, a number growing all the time, with the continuance of many wars and violent sociopolitical conflicts.

It seems to me, and some benefactors of refugees I have known, that the restrictions on refugee entry into the United States are far below what is needed and tolerable domestically, now as in the 1930s. One of the lessons I have learned about refugee resettlement is that the economic self-sufficiency and self-support of refugees, or motivations and prospects for the same, whether achieved by diligent entrepreneurship or professional competencies, has a definite positive impact on the willingness of Americans to bear the temporary burdens of welcoming more refugees into asylum in this country. (More on this later, where my mother succeeded and my father failed to demonstrate that desirable self-sufficiency by, respectively, successful and unsuccessful entrepreneurial activity.)

In this chapter, I am most concerned with three particular groups of refugees coming to the United States from particular regional contexts and times. I have had personal and professional involvement with each, although I am a member only of the first:

1. Central European (mostly German but also Austrian, Czechoslovakian, and Hungarian) Jewish refugees from Nazi persecution in the 1934–1945 pre- and WWII period
2. Russian-Jewish refugees from Communist persecution and discrimination (I had contacts with them from 1989 to 1999)
3. Liberian Mandingo refugees from persecution, killing, torture, and life-threatening violence in civil war

The first of these three groups of refugees, the one with which I am most familiar, was somewhat unrepresentative of refugees in general, particularly in recent times. At least so far as my mother's relatives and friends were concerned, they were, in the Germany of the 1920s and 1930s, privileged, affluent, educated, "cultured," well traveled, cosmopolitan in outlook, loyal to their native country, relatively conservative politically, and secular rather than orthodox in religion and behavior. German, Austrian, Czechoslovak, and Hungarian Jews in the nineteenth and early twentieth centuries had risen socioeconomically to widespread and respected middle- and upper-middle-class levels; they thought of themselves as loyal, respected, and well-integrated citizens of their national communities, and indeed proudly served their national governments before, during, and after WWI in both military and civil services with some distinction. Thus, their particular tragedy was that they considered themselves Germans (or Austrians, Czechs, or Hungarians) whose forebears happened to have practiced the Jewish religion. They themselves now rarely practiced it, and few could read or converse in Hebrew or Yiddish—they were as German as their Christian neighbors. They believed they were well accepted and even often befriended by their Protestant or Catholic neighbors, rather than Jews who happened to live in Europe and who were about to fall from socioeconomic prominence to become the abject victims of merciless persecution and genocide. The tragedy of these Central European Jews was that after their century-long progress to relative preeminence, they were the most poorly prepared psychologically for their fall and persecution. Many of them just could not believe what was happening to them, until it was too late.

My Story as a Young Refugee Entrepreneur, 1937–1950

My mother learned all the lessons of immigrant refugee entrepreneurship, but my father only some of them. At first, I was too young to understand them, but soon did. My own employment history, beginning at age seven with an unsuccessful shoeshine venture, rewarded my persistence and eagerness to work at whatever was available to me to earn money and a related degree of freedom of movement, entertainment, and purchases. Immediately after moving from two years of New Jersey boarding school back to Manhattan at age nine, all through age 12, I sought work in the neighborhood—first as a delivery boy earning 25 cents an hour plus occasional tips, then in a smelly butcher shop, then a step up to a grocery store, and next to an even cleaner, better-tipping work delivering flowers on holidays and medicines for two different drug stores. This might seem to raise the issue of immigrant child labor and how it can be both productively facilitated and humanely regulated, but I can say that between my mother's supervision and the demands of my school homework, I did not have to work excessively, unhealthily, or dangerously.

The importance of education as a means of advancement in income, power, and freedoms was always clear to me. Fascinated as I was from the start with technology—particularly the exciting WWII military technology of planes, ships, subs, tanks, and guns—I realized that I had to learn science and mathematics to be able to participate in technological progress and perhaps even contribute to its fight for freedom from oppression. It was drilled into me by my mother that I must always do my best, and I wanted to in more and more subjects. It was always understood that I should seek the highest and best degree of educational attainment by both schooling and informal means. In the eighth grade I took the competitive exams for high school, got into both Brooklyn Tech and Bronx Science, and chose the former. I read widely and deeply from about age 11 up. Tales of youthful entrepreneurial success, such as Tom Sawyer's, appealed.

From elementary school through high school, I learned that I could always get a job of some sort part time, and for weekends during the school year, full time. During summers, simply by making the effort in my home neighborhood or by asking adults in the area or by responding to signs in store windows, "Boy wanted." The worst-paying and endurance-proving full-time summer job I had was at age 13, as a waiter-bus-boy-dishwasher-lifeguard at a cheap refugee-run Catskill summer resort, arranged through friends of my mother's. I learned early that jobs arranged through friends of friends of parents were sometimes the absolute worst, mean, and exploitive. At this resort I worked 18 hours a day from 6 A. M. to midnight, with a couple of hours "off" for lifeguarding at the pool, and got yelled at by the owner's ill-tempered wife for not being quick enough. For pay I got room and board and occasional tips, and $10 a week.

The best-paying summer job I had—at age 14—was for $1 an hour filling and piling up 100-pound bags of ground malt for my mother's old school friend (another successfully entrepreneurial German-Jewish refugee from Munich) who had a malt and hops business in Patterson, NJ—a two-hour commute from my mother's apartment in the upper East side of Manhattan. I proved to myself I could stand eight hours of hard physical labor amid noxious dust, plus four hours of tedious commuting, and still have energy left to go partying and dancing a couple of working day nights a week.

The most pleasant part-time and brief summer jobs I had during the schools year were as an elevator operator in the Hotel Lexington and in

Number One Fifth Avenue. What I liked most about that easy work was ogling the often beautiful call girls and friendly bar girls while taking them up to their assignations on the upper floors.

The best-paying part-time job I had was shoveling snow in Manhattan, for the great sum of $20 for just four hours of hard work—but that was only during heavy snowfalls. By my seventeenth summer (1947), having graduated from high school with high honors in January and knowing that I would be admitted to MIT in September, I picked full-time jobs that both paid fairly well and would give me interesting experiences relevant to my chosen career as an engineer. In those seven months between high school and college I worked successively as a machinist, garage mechanic, plastics buffer, draftsman for the NY Central RR, as a composing room proof-press operator at the *New York Times*, and finally, closest to my intended career as an aeronautical engineer, as a loftsman and welder at Chance Vought Aircraft in Stratford, Conn. Six jobs in seven months were a reflection of my hunger for diverse experiences; my impatience as to getting some sort of paying work as soon as possible, given the need to save up for college; and my experience-based self-confidence that if I did not like one job I could always quickly find another.

A major step in my entrepreneurial career occurred during the 1947 MIT freshman term, when I was 18. It was a sandwich business, in which I and soon a classmate assistant bought cheap day-old bread from the old Ward Bakery near MIT; made ham, baloney, cheese, and peanut butter and jelly sandwiches in our crummy old wooden Building 22 dorm rooms in the late afternoon; and sold them in the building's study halls late into the night. My associate, Warren, and I were netting some $50 a week each, and soon I was able to realize one of my strongest ambitions, the purchase for $100 of my first car, a 11-year-old heap of a rusty 1936 Chevy sedan with flapping fenders and spavined upholstery, and a starter that had to be worked from under the hood, that nevertheless got my friends and me out to Wellesley and other places to pick up and take dates. Another freedom achieved through entrepreneurship.

My next entrepreneurial venture was not until three years later, as the writer of an undergraduate novel that I hoped to sell to the fiction editor at Houghton Mifflin, the wife of one of my favorite MIT professors, the philosopher Giorgio deSantillana. Writing and selling a book is certainly an entrepreneurial effort, whether successful or not.

All these entrepreneurial ventures (or adventuring, as they used to be called by the Elizabethans) are potentially accessible to new refugees and immigrants today, particularly if at least some of the most important lessons learned (presented later) can be learned and applied by them. Family cultural capital is in general probably more important than financial capital or even intellectual capital. The entrepreneurship forced on immigrants by a shortage of job opportunities—to make their own job if they cannot get one working for someone else—takes energy, initiative, persistence, self-confidence, pragmatic willingness to learn and begin modestly, and optimism about the future rewards of honest effort. These are the cultural characteristics of many ancient surviving cultures—Greek, Chinese, Cuban, Italian, Irish, Armenian, Caribbean, Jewish, Ibo, Tutsi, Mandingo, among others.

Entrepreneurial Lessons Learned by Refugees

What lessons were learned by my parents and me? The most important lessons learned by all three of my nuclear family and our German-Jewish refugee relatives and friends of the 1930s and 1940s are summed up next. Most also apply to the Russian-Jewish refugees of the last 20 years, and some also to the West African refugees of the last decades. While many of the lessons of the refugee experience also apply to the non-refugee immigrant experience, the emphasis here is on trying to help refugees from oppression or war adapt successfully to current conditions in the United States.

1. In an era of continuing or indeed increasing international and transnational violent conflict, civil wars, and intra-national violent communal conflict, if one is a member of a discriminated-against minority group, potentially vulnerable to persecution and to being subject to property confiscation, imprisonment, torture, and murder, as were the Jews in Nazi Germany in the 1930s, the Russian Jews in the Soviet Union, the Mandingos in 1990 Liberia, and the Iraqis in contemporary Iraq, *it is wise to prepare to escape the local oppression or war by quietly developing contacts with secure potential asylum countries.* It is wise to organize possible safe escape routes and forge relationships with the diplomats, business enterprises, or NGO's (e.g., Amnesty International) of one or more potential asylum countries where human rights are more respected. (A Liberian journalist, Hassan Bility, when in opposition to the bloodthirsty Charles Taylor, was wise enough to befriend the American consul and individuals in the international press, and managed thus to be flown out of Liberia to sanctuary in the United States.)

2. *To prepare to be permitted entry into and to survive in an asylum host country, it is helpful to develop a marketable skill that is readily transferable to, and in demand in, a potential host country.* For professionals these transferable skills are scientific, engineering, technical, and scholarly. Medicine

requires relicensing and possibly retraining, which some former Soviet physicians were willing to do and some not. In my acquaintance with six German-Jewish refugee physicians and one dentist in the 1930s in New York, I was impressed with their continuing their practices among fellow refugees, while retraining and qualifying for American licenses. Several Russian foreign medical graduates of my acquaintance in the 1990s were not equally willing to be retrained and relicensed, perhaps because their model of Russian medical practice was more at variance with the American model. Law and legal training is less transferable across national boundaries, being relatively unique to each country. My uncle Eric Midas, who held a doctorate in law from Heidelberg, was unable to practice law after coming to the United States; he opened a lackluster furniture store, and finally wound up selling law books.

3. *When seeking asylum from oppression in a host country such as the United States, refugees should not tarry long in an intermediate host country short of their intended destination.* The United States considers refugees who had been in residence for a year or more in a country other than that of their persecutors to have been granted asylum in that country; this spoils the opportunity for obtaining the ultimately desired asylum in the United States.

4. *When seeking asylum or even temporary residential status in a host country of refuge such as the United States, it is important to enter legally if at all possible* and avoid contaminating the case for more permanent immigration and residency status. Many refugees desperate to leave their oppressive native country and to come to the United States find their way to a nearby slightly safer one, and there may obtain a passport by illegal means, or otherwise obtain temporary entry into the United States on the basis of false information supplied to obtain an entry visa. Eventually, especially if they fail to renew their visa accurately and on schedule, the U.S. Immigration and Customs Enforcement (ICE) authorities catch up with them. They are brought before an immigration court, and there are ordered to depart the United States voluntarily within usually three months and to apply to reenter legally (if they can), on pain of being jailed, fined, and deported, and never having the right to reenter the United States for ten years or more.

5. It is, of course, very helpful for refugees and immigrants to *have friendly, supportive, and preferably well-established relatives or friends in the U.S. city of arrival* who are familiar with legal procedures, and employment and economical housing opportunities here. Even better, they may be able to provide a job in their own enterprises. However, to take fair advantage of any such "connections," and not burn out one's welcome with a supportive

American host, one must work humbly and hard to take whatever is offered and make the best of the opportunity.

6. If well-established relatives or good friends are not available, it may be necessary *as quickly as possible to befriend a sympathetic former countryman or co-linguist* in the resident neighborhood or at work, to learn how to get along in the new country. One way of doing that is through finding work at a large, well-established company, at whatever level is available, perhaps by first finding a countryman or co-religionist already established there, and requesting introductions to become a fellow employee.

7. Even affluent American relatives and friends, and especially nonaffluent ones, will *expect new refugees to quickly become self-supporting and as self-reliant as possible* by making their best efforts to find work, housing, and schooling for their children. *English language speaking, understanding, reading, and writing skills, however rudimentary at first, need to be developed as quickly as possible* for fully effective job search, housing search, child schooling, avoidance of exploitation by unscrupulous employers or landlords, and lawful behavior.

8. "Ignorance of the law is no excuse."

9. Particularly in times and locales of high unemployment, the new immigrant must be *ready to do whatever it takes to earn a living, however humble the work, and not expect to be employed at his/her previous higher level of skill or status.* Managers and engineers and affluent homemakers in the home country must be willing to adapt to new, reduced circumstances by taking whatever jobs are available, however humble. I have observed the failure of both my father (a former art and antique dealer in Germany) and several Russian refugees (former physicians, factory managers, and senior engineers) to accept more humble positions than they had achieved in their native countries, too proud and unwilling to start afresh, and thus destroying their opportunities to rise again. Pride in previous higher status can be a serious obstacle to a willingness to "start at the bottom," where there's usually plenty of room.

10. The new refugee has the best chance of learning to get along, picking up the language and the terminology and rules, acquiring new skills, and advancing into new roles by striving to *get a job in a large, diversified, and expanding business or industry,* rather than in a very small shop run by co-linguists that offers much less learning and advancement opportunities and job security. Tiny family enterprises also risk embroiling the new immigrant employee in family jealousies and conflicts, while at the same time requiring illegally long working hours of him and isolating him from

the more reasonable and diverse labor practices and opportunities within larger enterprises.

11. Non-English-speaking children must be prepared by parents to accept the initially harsh and confusing situation *of total immersion in all-English-speaking day care centers, kindergartens, and schools and be encouraged to learn English as quickly as possible. The children should also push their parents to speak English at home,* overcoming their tendency to speak nothing but the native, non-English language in the home. School children can often accelerate the acculturation of their parents.

12. A resilient, determined, realistically optimistic, self-reliant character that is paired with an egalitarian, meritocratic attitude helps a lot in the new immigrant/refugee's adjustment to his/her new status. An egalitarian outlook offers hope of rising meritocratically. A meritocratic outlook offers hope of strong effort being justly rewarded. Together, they facilitate growth.

13. Family solidarity even if with only one relative—in my case, with my mother—is an important asset, even if the whole family is poor. There is cultural capital, collective experiential knowledge, and there are positive role models and negative cautionary tales to be learned and adapted to the new situation. There are good reasons for immigrants clustering together in family groups and, if without family, in age or linguistic or ethnic affinity groups. There is both strength and wisdom in community however small and estranged from home.

14. "Dress for success," or try to fit in with the surrounding age- and gender-appropriate roles, manners, and mores. It helps to defray the not uncommon xenophobic biases of employers, associates, and landlords.

15. Contact as soon as possible co-lingual, co-ethnic legal and medical assistance. Acquaintance with a lawyer, law firm, or the legal department of a major employer is important to being informed about immigrant/refugee rights and responsibilities and the best means of becoming a legal resident. This is particularly important in the current xenophobic atmosphere of suspicion of undocumented aliens. Acquaintance of a co-lingual, co-ethnic physician or nurse is particularly important if small children or medically challenged adults are among the new immigrants/refugees. I was fortunate in both respects, my family being acquainted with many fellow refugee physicians (who in fact made house calls even before they could get relicensed). In the absence of often crucial connections, new immigrants should be guided to legal aid and medical aid organizations as soon as possible, with co-lingual capacity or, lacking that, promptly guided to English language training.

16. Young refugees/immigrants will need financial help to establish their enterprises in the United States. But before asking anyone for money to finance a venture, the young refugee/immigrant must conceive and write down his/her business concept. The basic business concept describes very briefly what service/product will be sold to what customers in what markets at what price and what cost, what are the prospects for expanded profitable sales (priced above cost), and why the young refugee/immigrant is able to deliver this service/product better than competitors and knows something about its market. Ideally, the basic business concept will have an innovative idea for producing an unconventional product for a conventional market or a conventional product for a new market, or making the product/service-market connection in a new way superior to competitors. Going to banks or businesses to borrow money to start a business is useless until you can demonstrate that you know something and have skills that are useful to someone—that is, you must have a job. Get the job before seeking financing, and obviously get the job in an organization that interests you and can make best use of your cultural capital and increasingly skilled interests. If you are uneducated, the first thing to do is to obtain those rudimentary language, arithmetic, and social skills essential to get a job as described earlier. On the job, you can "learn while you earn" and keep advancing in skills, status, and self-confidence, enough to create a new and promising business concept and either self-finance it from the savings of your jobs (and those of your friends and relatives who would be willing to invest in you) or finance it from a corporate "angel" who likes your idea and has confidence in you.

17. There is the enormous, possibly decisive, impact of cultural capital on the future socioeconomic success of immigrant/refugee entrepreneurs, certainly in the case of Central European bourgeois refugees from Nazi Germany or Austria or other countries. The content of this cultural capital for me, personally, was the object-embodied love, possession, and appreciation of classic and modern literature, poetry, fine arts, and antiques; professional medical, legal, business, and scholarly skills; and even some military combat infantry experience and sophisticated travel experience embodied by my immediate and extended family. (My mother brought from Germany her collection of German classics—Goethe, Schiller, Hauptmann, Mann, Heine, etc.—and even a 1923 edition of the complete works of Shakespeare as well as many fine antiques and paintings.)

To be sure, a problem with these "lessons learned" is that they are transferable to the current generation of African, Asian, and Latin American immigrant/refugee entrepreneurs only in part. Thus, with respect to, say, some Nigerian refugees from the Biafra war or poor Mexican immigrants, their cultural capital may take quite different forms than a knowledge of their native literature, art, or enterprises. A better knowledge of the cultural capital brought by the major current and coming immigrant/refugee groups would provide a better basis for assisting them to connect their own cultural capital inheritances with contemporary American society and economy, and apply those connections for their own entrepreneurial success. For example, I doubt if classic Spanish and contemporary Mexican literature make up much, if any, of the cultural capital of poor Mexican migrant-worker immigrants. What does? We should find out.

Conclusion

I am quite aware that I was fortunate to have had the cultural capital and role models, positive and negative, to succeed. But it is also significant that my role models—the successful ones (my maternal grandfather and remote American relations); marginally successful ones (my aunt, uncle, and mother); and quite unsuccessful ones (my father)—were all entrepreneurs. To some important degree, I believe that any entrepreneurial success I have had was overdetermined by that. Still, I believe that for current young immigrants/refugees with the taste for it, many of the "lessons learned" will apply, and helping them to become entrepreneurs will be good for them and for our country.

Part III

Recent and Current Waves of Young Immigrants and Refugees

This part focuses on the current immigration streams and on the issues faced by those young immigrants and refugees as well as American society. In contrast with much of the "old immigration" and also with early refugee movements, such as the Second Wave discussed in Part II, the "new immigration" and the more recent refugee streams have predominantly been nonwhite. This heightens the importance of the racial and ethnic dimensions in the situation of young immigrants and refugees. The chapters by Kasinitz and by Halter and Johnson most explicitly address this topic (as well as the one by Alba and Tsao that appears in Part VI).

In a chronological sense, the first chapter of this part, Steven J. Gold's study of Cold War refugees from the Soviet Union and Vietnam, links the World War II immigration that was the focus of Part II with the more current immigration streams. It shows how strongly political considerations shaped American policies regarding refugees. Gold goes on to compare the Cold War refugees with post – Cold War refugees from Africa, especially the Sudanese Lost Boys.

The second chapter in Part III, by Mary Waters, reports about the New York Second Generation Study, a study of children of immigrants, living in New York City, and conveys an optimistic message about the adjustment of these children. Contrary to many scholars' expectations of a second generation "downward assimilation," this study found that the children of immigrants were, on an average, doing better than their immigrant parents. Moreover, the young immigrants of all ethnic backgrounds did better than native contemporaries of comparative ethnic and racial descent.

In the following chapter, Philip Kasinitz turns our attention to how the receiving society views the arriving young immigrants and refugees and how it attempts to aid them. He focuses specifically on the reception of Russian Jewish refugees and on the consequences of affirmative action policies. These young refugees from the Soviet Union have received a good amount of support from established and well-networked aid organizations. Furthermore, a cultural shift toward diversity and societal and policy changes brought about by the civil rights movement have also helped young immigrants, especially those classified as minorities.

Marilyn Halter and Violet Johnson focus on young West African immigrants and refugees. They describe these young immigrants' "psyche for success" as rising from an amalgamation of homeland traditions, collective memory, and social capital. In addition, the authors explore the complicated identity issues that challenge the hold of the parents' immigrant subcultures on their children.

The legal dimension of immigration is paramount in the current public debate. The chapter by Frank D. Bean, Susan K. Brown, Mark A. Leach, and James Bachmeier addresses a specific aspect that pertains to young immigrants; it examines the social consequences of parental legal status for young immigrants from Mexico. The authors found that parents' legalization as well as their naturalization had multiple beneficial outcomes for children, such as lower high school drop-out rates, higher college completion rates, more prestigious jobs, and higher earnings.

6

After the Cold War: Comparing Soviet Jewish and Vietnamese Youth in the 1980s to Today's Young Refugees

Steven J. Gold

Introduction

From the late 1940s until the fall of the Soviet Union in 1991, the Cold War provided the central frame for U.S. refugee policy. During this time, the United States was embroiled in political and ideological conflicts with the Communist Bloc. Accordingly, it sought to demonstrate its superiority by extending refuge to those who fled Communist repression. Cold War refugees represented specific elements of their countries of origin. They included prerevolutionary elites who were opposed to the new regime and could use their resources to resist it. Similarly, entrepreneurial minorities were often ejected because of their association with capitalism, their "dubious loyalty," and their extra-national affiliations. Such groups were subject to retribution because—deservedly or not—they were blamed for prerevolutionary abuses and used as scapegoats for the failings of the new régime. Faced with such treatment, many had no choice but to take flight.

As a consequence, Cold War era refugees often came from the upper echelons of their home countries and were characterized by high levels of skill, education, and familiarity with Western life. In granting them refuge, the United States not only discredited the regimes that sent them off, but further availed itself of a population endowed with significant human resources. In a strangely symbiotic relationship, revolutionary societies "benefited" through the ejection of skilled classes while providing capitalist states with refugees whose resettlement offered both political and economic rewards.

First wave Cubans, pre-1977 Vietnamese, and Jews from the Soviet Union epitomize this phenomenon. Their speedy adjustment to the United States was further enhanced by relatively nontraumatic preexit conditions, travel abroad under good circumstances, and access to high-quality resettlement services. This experience was especially conducive to the adaptation of the younger members of these populations, who have done quite well in the United States.

Of course, not all refugees entering the United States during the 1970s–1990s were skilled, educated, and relatively free from trauma. Cambodians, Hmongs, Ethiopians, and others lacked education and skill and were subject to extreme violence prior to exit, traumatic flight, and long stays in primitive refugee camps before their eventual resettlement. Once in the country, they confronted culture shock, unemployment, poverty, discrimination, and resettlement in declining neighborhoods (Hein 2006) (see table 6.1).

Since the end of the Cold War, the criteria for selecting refugees and, as a consequence, the characteristics of refugees themselves have changed. Describing the "US Refugee Program in Transition," David Martin (2005) asserts:

> Largely gone are the massive, steady, and more predictably manageable programs that had dominated US admissions since the passage of the Refugee Act of 1980—the Indochinese and Soviet programs, followed for a few years by programs for those fleeing the former Yugoslavia. For the future, refugee admissions will be characterized by the combination of many smaller-scale resettlement programs, mostly originating in difficult locations that will shift from year to year, each presenting significant and distinct policy challenges.

Despite redirected priorities, considerable numbers continue to arrive as refugees and asylees from Cold War era sources, including Russia, the Ukraine, Laos, and Cuba. Others enter from Iran, which, while not part of Cold War geopolitics, has been a source of refugees since the Cold War era. And Iranians, like other Cold War era refugees, are characterized by generally high levels of education, skill, and urban experience (Gold and Bozorgmehr 2007).

While political concerns continue to shape the admission of refugees, the highest priority for admittance is now reserved for groups identified by the UN High Commissioner for Refugees (UNHCR) as being in urgent need of resettlement (Patrick 2004). These include persons from Somalia, Liberia, and Sudan.

Table 6.1 Summary of refugee admissions for FY 1975–FY 2005

Fiscal Year	Africa	East Asia	Eastern Europe	Soviet Union	Latin America	Near East Asia
1975	0	135,000	1,947	6,211	3,000	0
1976	0	15,000	1,756	7,450	3,000	0
1977	0	7,000	1,755	8,191	3,000	0
1978	0	20,574	2,245	10,688	3,000	0
1979	0	76,521	3,393	24,449	7,000	0
1980	955	163,799	5,025	28,444	6,662	2,231
1981	2,119	131,139	6,704	13,444	2,017	3,829
1982	3,412	73,755	11,109	2,760	580	6,480
1983	2,645	39,245	11,867	1,342	691	5,428
1984	2,749	51,978	10,096	721	150	4,699
1985	1,951	49,962	9,233	623	151	5,784
1986	1,322	45,482	8,503	799	131	5,909
1987	1,990	40,099	8,396	3,699	323	10,021
1988	1,593	35,371	7,510	20,411	2,497	8,368
1989	1,902	45,722	8,752	39,602	2,604	6,938
1990	3,453	51,598	6,094	50,628	2,305	4,979
1991	4,420	53,522	6,837	39,226	2,253	5,342
1992	5,470	51,899	2,915	61,397	3,065	6,903
1993	6,967	49,817	2,582	48,773	4,071	6,987
1994	5,860	43,564	7,707	43,854	6,156	5,840
1995	4,827	36,987	10,070	35,951	7,629	4,510
1996	7,604	19,321	12,145	29,816	3,550	3,967
1997	6,065	8,594	21,401	27,331	2,996	4,101
1998	6,887	10,854	30,842	23,557	1,627	3,313
1999	13,043	10,206	24,497	17,410	2,110	4,098
2000	17,561	4,561	22,561	15,103	3,232	10,129
2001	19,021	3,725	15,777	15,748	2,973	12,060
2002	2,548	3,525	5,439	9,963	1,933	3,702
2003	10,717	1,724	2,525	8,744	452	4,260
2004	29,125	8,079	489	8,765	3,556	2,854
2005	20,749	12,071	141	11,175	6,700	2,977
1975–2004 Grand Total	184,955	1,300,694	270,313	616,275	89,414	145,709

Note: This chart does not include an additional 8,214 refugees admitted between FY 1988 and FY 1993 under the private sector initiative (PSI) or the 14,161 Kosovar refugees admitted in FY 1999. Numbers listed for Latin America exclude Cuban entrants. Since 1960, 817,837 Cubans have entered the United States. Most of these Cubans came as refugees, even though the specific laws that permitted their entry did not always refer to them as such (Perez 2007: table 2, Haines 2007).
Source: Office of Refugee Resettlement, Report to Congress, 2007.

Table 6.2 Selected nationalities, percent in occupational categories

Nationality	Manager/ Professional	Technical/ Sales/Admin	Service	Production/ Craft	Unemployed	Self-employed
Vietnam	23.3	22	17.8	36.6	0.6	11.1
Russia	45.2	27.8	8.7	17.3	0.8	12
Somalia	16.6	23.7	15.3	39.7	3.8	4.9
Sudan	28.2	27.1	13.1	29.8	1.9	4.5

Source: 2000 (5 percent PUMs).

Table 6.3 Selected nationalities, social characteristics

Nationality	Less than High School %	Median Family Income $	Median Personal Income $	% Household Linguistic Isolation *	Median Age in Years
Vietnam	35.7	50,000	23,150	43.3	36
Russia	7.8	45,200	29,300	34.4	39
Somalia	36.1	17,300	16,000	45	25
Sudan	17.4	23,500	19,300	30.8	30

* All members of the household age 14 and above have difficulty with English
Source: 2000 (5 percent PUMs).

In sum, the end of the Cold War context for U.S. refugee policy means that the experience of refugee youth from the post – Cold War era will be unlike that of those who entered the United States from the 1970s through the early 1990s. The economic and cultural climate of the United States—characterized by benefit-limiting welfare reform, post – 9–11 restrictions on migrants, and the political mobilization of anti-immigrant sentiments—appears to have made the country less welcoming for refugees and migrants than during the 1970s and 1980s. However, the major reason that today's refugee youth are unlikely to match the record of those who arrived in the recent past has less to do with the receptions refugees receive in U.S. society and more to do with their origins (see tables 6.2 and 6.3).

Cold War Refugees

Many young Soviet Jewish and Vietnamese refugees of the 1970s through the 1990s achieved impressive levels of mobility. A major reason for this is that their families were endowed with significant resources, notably education and urban experience. In addition, these groups maintained family structures and practices that allowed them to effectively support their children shortly after arrival.

Soviet Jewish and Vietnamese refugees tended to settle in locations where there was a demand for their occupational skills. Fortuitously, they entered such locations prior to the recent price inflation in residential real estate. Accordingly, many were able to rent an apartment or purchase a condominium or house in a middle-class neighborhood where jobs, quality schools, and other services facilitated the family's adaptation. These populations benefited from more generous resettlement benefits than are available to contemporary refugees. (In 1980, refugees could receive up to three years of benefits. Currently refugees receive a maximum of eight months.) Finally, both groups received services from co-ethnic resettlement organizations.

Few contemporary refugees are characterized by this combination of extensive resources and fortuitous circumstances of settlement. Recent reports indicate that the largest numbers of contemporary refugees (over a third admitted during 2004 and 2005) are from Somalia and Laos. As such, they are less educated and more traumatized than Vietnamese and especially Soviet Jews (Jeffrey 2006). In addition, resettlement benefits have been reduced, whereas residential real estate markets in major cities have become prohibitively expensive for recent refugees. Consequently, the accomplishments of recent arrivals may be more modest than was the case for some of those who entered the United States between the mid-1970s and the early 1990s. In the section that follows, I summarize the social characteristics of Soviet Jewish and Vietnamese refugees in the United States, and offer some brief comparisons with recent refugee populations.

Soviet Jewish Refugees in the United States

Many of the social characteristics and conditions of settlement surrounding the Soviet Jews who entered the United States between the late 1970s and the early 1990s were as close to optimal as one could expect for the adaptation of refugees and their children to the United States. This population was highly educated and hailed from the urban, European parts of the former Soviet Union. Prior to exit, Jews had been subject to religious and ethnic discrimination, denied religious freedom, and restricted in opportunities for jobs and education. Nevertheless, they had achieved a high educational profile. Moreover, their pre-emigration lives were not nearly as arduous or risk filled as those of other refugee groups who encountered war, ethnic cleansing, deportation, and organized violence prior to their flight (Gold 1995).

In addition, the family strategies that Soviet Jews established to cope with life in the urban Soviet Union proved to be highly applicable to life in the United States. These included a small number of children, multigenerational, intact families; relatively few years between generations; and high rates of marriage (Gold 1995). These enabled families to devote extensive resources to each child in such a way as to assist with adjustment.

Russians had access to a package of benefits, including health care, job and language training, housing, public assistance, and legal resident status. As Jews, they could obtain services provided by the American Jewish community, including Jewish Family Service, Jewish Community Centers, Jewish retirement homes, Jewish schools and synagogues, and other programming. Jews as well as Armenians and other ethno-religiously defined populations of refugees from the former Soviet Union also have access to networks of native-born co-ethnics that often aid their resettlement by providing guidance, jobs, and various services (see Kasinitz in this volume).

Russians who have entered the United States since 1970 are characterized by very high levels of education—often in technical and professional fields. Reflecting those high levels of education, recent Russian immigrants experience rapid economic mobility. According to the 2000 census (5 percent PUMs), Soviet-born men in the labor force age 25–64 were making a median income of $38,000 in 1999. For purposes of comparison, the median income for all foreign-born men age 25–64 in the United States who were in the labor force was $27,000 in 1999. According to the 2000 census, Soviet-born women in the labor force age 25–64 were making a median income of $24,500 in 1999, while the median income for all foreign-born women age 25–64 in the United States who were in the labor force was $19,400 in 1999. With the exception of the sizeable fraction of Russians who are elderly and hence not employed in the United States (40.4 percent are 45 or older), many have been able to join the American middle class.

Russian refugees and their descendants find themselves among the upper echelon of all immigrants and refugees in the United States due to their largely European origins, high levels of education, legal status, connections to established co-ethnics, and white skin. While many immigrant and refugee groups become racial minorities upon entry into the United States, Jewish émigrés from the former Soviet Union often lose their public Jewish identity since non-Jewish Americans simply see them as white foreigners and American Jews find them lacking in religious knowledge. An émigré interviewed by sociologist Paul Ritterband (1997: 332) in New York commented on his loss of minority status: "Here I feel less a Jew than in Russia . . . I live calmly and nobody bothers me . . . Nobody tells me I'm Jewish."

Because Russian immigrants have relatively high rates of self-employment, their population has created enclaves like those in West Hollywood and Brighton Beach that feature numerous ethnically oriented shops, restaurants, service providers, and media industries that present a venue for socializing and identity reconstruction. The sum total of these many positive characteristics has been a positive experience of resettlement for a significant proportion of Soviet Jewish youth in the United States.

Education of Russian Refugee Youth

School-aged Russian immigrants who have entered the United States since the 1970s generally have a strong educational background and do well in American schools. For example, in a 1991 comparison of the 12 largest immigrant groups attending New York City public schools, grades 3–12, who had been in the country three years or less, students from the former Soviet Union ranked first in reading scores, second in math, and fifth in English. Their reading and math scores were much higher than the average for all students, including the native born. In addition, their mean increase in score over the previous year was the highest of all groups in both reading and English and among the highest in math (New York City Public Schools 1991). According to the 2000 census, 53 percent of Russians in the United States were in high-status professions (40 percent of the U.S.-born population has such occupations) (Portes and Rumbaut, 2006: 221).

Vietnamese Refugees

In contrast to refugees from the former Soviet Union, the Vietnamese have had a multifaceted experience. As a group, they have done relatively well. Yet their community is quite diverse in terms of background, experience of exit, and tenure and fate in the United States. Their numbers incorporate several vintages characterized by distinct patterns of adaptation. These include the first wave, boat people, Amerasians, and re-education camp survivors.

The first group of Vietnamese refugees, numbering about 175,000, avoided many of the most traumatic elements of flight from Vietnam. Reaching the United States between 1975 and 1977, most were U.S. employees and/or members of the South Vietnamese military and government (and their families). Evacuated by American forces, first wavers avoided life under the new regime in their home country and spent only a short time in refugee camps before being resettled in the United States, often with immediate families intact. Drawing upon their skills, education, competence in English, familiarity with Western culture, and extended families, many adjusted rapidly (Gold 1992).

The second group of Vietnamese began to enter the United States following the outbreak of the Vietnam-China conflict of 1978. These refugees were generally of lower-class standing than first wave refugees and included both Vietnamese and ethnic Chinese populations. They had lived for three or more years under Communism, sometimes laboring in re-education camps or remote "new economic zones," before fleeing Vietnam.

Their exit, involving open sea voyages in leaky, overcrowded boats or long journeys on foot across revolution-torn Cambodia to Thailand, was subject to attacks by pirates and military forces. Up to half of these refugees perished in flight. Having crossed the high seas, boat people then languished months or even years in the overcrowded refugee camps of Thailand, Malaysia, Indonesia, the Philippines, or Hong Kong before entering the United States (Gold 1992).

Due to the dangers of escape, far more young men left Vietnam as boat people than did women, children, or the elderly. For example, in 1984 there were approximately 204,000 male Southeast Asian refugees in the United States between the ages of 12 and 44, but only 156,000 females of the same age group. Since that time, the sex ratio has moderated. As of 2000, Vietnamese women outnumbered men (United States Bureau of the Census 2000).

In addition, post-1978 refugees experienced financial troubles more severe than those encountered by earlier arriving Vietnamese. While Vietnamese refugees coming to the United States in 1975 averaged 9.5 years of education, and two-thirds knew some English upon arrival, those entering between 1980 and 1982 had an average of seven years of education, and half had no competence in English. Disadvantaged by lower levels of education and English language ability, and contending with the depressed U.S. economy of the early 1980s, boat refugees faced difficulty in finding their first American jobs and had to deal with shrinking government benefits, as well.

Among the post-1978 refugees from Vietnam (as well as Cambodia and Laos) was a sizeable population of ethnic Chinese. According to the 2000 census, 11.9 percent of Vietnamese refugees in the United States claimed a "Chinese" identity. Constituting an entrepreneurial class, this group has a long history but marginal status in Southeast Asia. Prior to 1975, Saigon's Chinatown of Cho Lon (literally, big market) neighborhood was superseded only by Singapore as the largest Chinese settlement outside of China.

Relying on past experience and on capital and goods from overseas Chinese communities, a number of ethnic Chinese have been able to reestablish their role as entrepreneurs in the United States, both in established Chinatowns as well as in new Southeast Asian enclaves like Orange County's Little Saigon. While some of the most financially successful members of the Southeast Asian population in the United States are of ethnic Chinese origins, collectively they tend to be disadvantaged when compared with the ethnic Vietnamese. Lacking a Western-style education and sometimes subject to discrimination from ethnic Vietnamese in the United States, their economic adjustment has been slower than that of the ethnic Vietnamese, and marked by distinct patterns of education, residence, and occupational distribution (Gold 1994).

Since the late 1980s, the U.S. government has established special programs for Amerasians and re-education camp survivors, two groups of Vietnamese who have been subject to particularly harsh treatment in their home country. Amerasians are youth whose fathers were American servicemen, while re-education camp survivors are former political prisoners who generally held high positions in the South Vietnamese military or government. In contemporary Vietnam, both groups are socially marginal and encounter both popular and official discrimination. Because they have an American parent, Amerasians are U.S. citizens, but due to their circumstances, they are also entitled to the full package of refugee benefits (Gold 1999).

Economic Status and Occupational Profile

First wave Vietnamese—as an elite group whose numbers include professionals and other high-level workers—matched the average earnings of the U.S. population by 1987. As of 2000, labor force participation rates (the fraction of those working or looking for work) for all Vietnamese—70.8 percent—exceeded that of all foreign born (60.6 percent) and all native born (63.9 percent) in the United States (Portes and Rumbaut 2006: 78). Of those employed, 36.6 percent of Vietnamese age 25–64 held blue-collar jobs. By 1999, the Vietnamese median household income was $53,993, exceeding that of the native born by $4,500 (Portes and Rumbaut 2006: 89).

Various scholarly and journalistic accounts suggest that in an attempt to compensate for economic hardships, Vietnamese refugees, along with other Southeast Asians, are relatively active in the informal economy, supplementing their regular incomes with "off the books" pay earned in businesses, swap meet sales, home-based production, et cetera (Arax 1987).

Considering that a disproportionate number of Southeast Asians live in California and other West Coast settings associated with a very high cost of living, their economic plight certainly is a difficult one.

Family Adaptation

Despite their relatively recent arrival and lack of familiarity with American life and the English language, Vietnamese refugees have been able to keep families and communities intact, and have developed generally positive ways of retaining valued traditions while coping with the opportunities and challenges presented by American society. Notable among these have been increased gender equality and opportunities for women and girls, which, while threatening to some traditional family patterns, yield economic and child-rearing benefits for entire families (Zhou and Bankston 1998).

Several studies have shown that the structure of Vietnamese American families is a vital source of social, economic, and psychological support for the population. For example, partly because of their positive health practices, Southeast Asians have among the lowest infant mortality rates of any ethnically defined population in the country. This is found even though they rank low in income, educational level, and English ability—indicators commonly associated with elevated rates of infant mortality (Portes and Rumbaut 2006: 192).

In fact, the achievements of Vietnamese and (other recent Asian immigrants) have been so broadly publicized that several critics accuse the authors of such works of creating a model minority myth of Asian immigrant success that diverts attention from the very real needs that this population has, while simultaneously making invidious comparisons with native-born minorities, who have less impressive records of accomplishment (Kibria 2002).

Even though Vietnamese families did well, on an average, there are exceptions. Because of economic problems, difficulties in adjusting to American society, and resettlement in declining inner-city communities, a fraction of alienated Southeast Asian youth have become involved in gangs and other criminal activities. Because of the frequency of home invasions and crimes of extortion, a 1989 *LA Times* survey found that fear of co-ethnic gangs and crime was the greatest concern of Orange County Vietnamese (Emmons and Reyes 1989). Reflecting the community's frustration with the threat of co-ethnic gangs, in 1993, Vietnam-born Westminister, California, city councilman Tony Lam (who made controlling crime a cornerstone of his campaign) called for the deportation of any non-citizen convicted of a serious offense (Gold 1999).

Community Formation

While the subgroups that make up the broader Vietnamese population share common perspectives resulting from their opposition to Communism and experience as refugees, they are also diverse in their political, religious, and ideological outlooks. The community reveals a mélange of contentious factions, vying for community leadership and addressing a variety of issues, interests, and constituencies. Bases of communal activism include veterans' associations, business development groups, resettlement agencies, and religious leaders, many of which publish their own newspaper or magazine. The ethnic Chinese already understand something about providing mutual assistance, having been able to draw upon their premigration experience as an organized ethnic minority in Southeast Asia. Relying upon their involvement in small business and their links with co-ethnics in the United States and overseas, they have made significant progress toward rebuilding organized communities in the new setting.

Vietnamese American communities have been unified by active media industries producing newspapers, magazines, and radio and cable television programs as well as by frequent festivals. Ethnic business districts offer a variety of services, restaurants, and coffee shops, and all kinds of wholesale and retail establishments, while an extensive recording industry produces many styles of music (ranging from traditional to MTV influenced). Such communities are valuable assets for maintaining country of origin – based outlooks among the refugee population; they also act as a familiar social space that can provide solace and a sense of home in an otherwise unfamiliar society. In sum, Vietnamese have made significant progress in the United States, but continue to face challenges as well.

The Education of Vietnamese American Youth

One of the most positive aspects of the Southeast Asian refugee experience has been the ability of this group's members to do relatively well in school despite the trauma of being refugees, their cultural and linguistic unfamiliarity with the American context, and the fact that many attend poorly funded urban schools (Caplan et al. 1992). However, while American-educated Vietnamese youths have made marked progress, the educational attainment of the Vietnamese population as a whole remains well below the national average and manifests a bifurcated pattern. As of 2000, just under 36 percent of Vietnamese, age 25–64, had less than a high school education, while 20 percent were college graduates. This can be contrasted with

23 percent of individuals with less than a high school education and 20 percent college graduates within the general U.S. population. As of 2000, 30 percent of all Vietnamese were in high-status professions, 10 percent less than the corresponding fraction among the native born (Portes and Rumbaut 2006: 221).

Post – Cold War Refugees

A sizeable number of refugees continue to enter the United States from countries associated with Cold War era conflicts, including Russia, Ukraine, Cuba, Vietnam, and Laos. However, many of the largest sources of refugees are now in Africa, notably countries such as Somalia, Sudan, and Liberia. More traumatized, less educated, separated from relatives, and having lived in refugee camps for years, the new arrivals are unlikely to experience as rapid and smooth adjustment as Soviet Jews and Vietnamese. Further, as the number of refugee arrivals have reduced, so too has the availability of data on the social characteristics and resettlement of recent populations. This makes it difficult for researchers and policy makers to evaluate their adjustment. Yet available information does offer a basis of hope.

Sudanese Lost Boys

To explore the experience of a post – Cold War refugee group, I summarize the experience of the Sudanese "Lost Boys." This group is both relatively well documented and offers a striking contrast—in terms of numbers, background characteristics, and patterns of resettlement—with the experience of Soviet Jews and Vietnamese. Sudanese "Lost Boys" (and a small number of girls) are unaccompanied minors and young adults who were separated from their parents during the 1983 Sudanese Civil War. Between the ages of 4 and 12, they fled to Ethiopia and later Kenya, and lived with minimal support and supervision for nearly a decade until they were resettled. Some 3,300 were hosted by the United States in 2001. (Note the small number as compared with the 1.3 million Southeast Asian and 600,000 Soviet refugees resettled in the United States from 1975 to 2003 (ORR 2003) (see table 6.1).

The majority of these young Sudanese display some symptoms of post-traumatic stress disorder, and many confront problems associated with adaptation and being racialized in the United States. (Among their woes is an unfortunate label that revives the Jim Crow tradition of identifying black men as boys.) Yet, these youth also reveal many promising signs of successful adjustment. These include ready access to social support, a desire for education, school attendance, completion of high school, and enrollment in a two- or four-year college (ORR 2003, Bates et al. 2005, Luster et al. 2006).

According to the Office of Refugee Resettlement, as of 2003, "Lost Boys" over 16 years of age had employment rates considerably higher than the entire population of U.S. refugees (86 percent for Lost Boys versus 55 percent for non – Lost Boy refugees). In addition, their labor force participation rate of 91 percent was 30 points higher than that of the five-year U.S. refugee population and 25 percent higher than the non-refugee U.S. population. They have made good progress in acquiring education, finding employment, becoming financially independent, and mastering English. In 2002, of those not working, 90 percent reported that this was due to problems with English. A year later, less than 10 percent cited the same reason for not working (ORR 2003, figure 1). Males, who are the overwhelming majority of this population, tend to do much better in acquiring work and learning English than females (ORR 2003).

Scholars attribute their positive adaptation to the assistance they receive from fellow refugees and host families and communities, to good quality resettlement services, and to culturally appropriate education and counseling programs. They also credit resilience and coping skills acquired during their flight from Sudan and their long residence in refugee camps prior to entering the United States with fostering positive outcomes (Luster et al. 2005, 2006).

An advantage enjoyed by recent refugees as compared with earlier arrivals is that U.S. society is now more capable of assisting refugees than it was in previous decades. For example, the fraction of the foreign born in the nation has increased from 6.2 percent in 1980 to 11.1 percent in 2000, making recent immigrants appear less exotic (Camarota 2002). At the same time, the infrastructure and sociocultural environment for dealing with recent arrivals who are not party to U.S. culture and English—ranging from service agencies and schools to health practitioners and ethnic media and food stores—have expanded considerably, not only in gateway cities like Los Angeles and Miami, but also in smaller municipalities in the Northeast, Midwest, and South, where increasing numbers of refugees now resettle (Layton 2006).

Especially since the early 1990s, crime rates in many urban neighborhoods—the locations where refugees often reside—have reduced, making points of settlement less dangerous than was the case in an earlier era. Finally, while the majority of refugees continue to locate in states and cities where a large refugee presence is already established, growing numbers of refugees are being resettled in small communities where real estate is more affordable, social services and jobs more available, and schools less crowded (ORR 2003).

Conclusions

The political context of the Cold War resulted in unusually skilled refugees settling in the United States. Since the end of the Cold War era, a refugee program directed, at least to some extent, toward heavily traumatized groups has taken its place. Lacking certain advantages and the resources of Russian and Vietnamese refugees, recent arrivals and their children will likely face a more difficult and slower resettlement process than the earlier groups. However, given the very trying circumstances that they faced prior to their exit, it is encouraging that they are offered the chance to build new lives in an environment that promises to afford them opportunities that would have been otherwise unavailable.

Such populations may yield fewer valedictorians, Ivy League graduates, and major contributors to the professions and the arts—at least in the short term. However, available research suggests that recently arrived refugees do relatively well, especially when we consider the many obstacles that they have had to overcome prior to their arrival. We can celebrate these accomplishments and also take some pride that U.S. refugee policy now reflects individuals' needs for resettlement as well as geopolitical realities.

Writing just after the conclusion of the Cold War era, Stephen Castles and Mark J. Miller (1993: 145–146) noted that Africa was the location of the largest number of refugees, yet very few of these were resettled in the West. The fact that the largest refugee admissions ceilings are now devoted to Africans, and that persons from that continent accounted for 23–44 percent of all refugees admitted to the United States in the fiscal years 2003–2005, is certainly an encouraging trend (Jeffrey 2006: 2).

Resettlement staff should not be discouraged by the difficulties encountered by contemporary refugees. These are to be expected. Rather, they should be gratified that persons especially in need now account for a significant fraction of entrants. While the challenges involved in helping them build lives in the United States are considerable, such efforts provide valued opportunities to those sorely in need.

7

Second Generation Advantages: Recasting the Debate*

Mary C. Waters

I think I have benefited from being Colombian, from being Hispanic. It's the best of two worlds. You know that expression? Like being able to still keep and appreciate those things in my culture that I enjoy and that I think are beautiful and at the same time being able to change those things which I think are bad. (Second Generation Colombian woman, age 23, New York City)

In 1995 I began a collaborative effort with Philip Kasinitz, John Mollenkopf, and Jennifer Holdaway to study the children of immigrants in young adulthood in New York City.[1] Like many other scholars at that time, we framed our study around a concern with "second generation decline" (Gans 1992). We were worried that as the children of recent immigrants became Americans they were at risk of what has been termed "downward assimilation." We feared that a significant proportion would earn less than their immigrant parents, have lower educational attainment, and have lower levels of civic participation in their new society. Furthermore, we also suspected that upwardly mobile immigrants would achieve their success in large part by remaining tied to the ethnic communities and economic niches of their parents. In contemporary America, we speculated, the most successful immigrant families might be the ones who kept large parts of "mainstream" American culture at bay.

More than a decade later, the results of the New York Second Generation Study suggest that the debate should be reframed, as the title of this volume suggests, toward explaining and documenting successful second generation outcomes. While we found examples of both marked downward mobility and success achieved through remaining in the ethnic enclave, neither turned out to be very common. On the whole, we found that second and 1.5 generation New Yorkers are doing better than their immigrant parents. Among the Chinese and Russian Jews, rapid mobility is, in fact, the norm. Some of this can be attributed to their parents' premigration class backgrounds and "hidden" human capital—but, particularly among the Chinese, the upward mobility among those from working-class backgrounds with very low parental human capital is stunning.

Not surprisingly, among those groups who are "racialized" as "black" or "Hispanic" in the U.S. context, the record is more mixed. Race and racial discrimination remain significant factors in shaping their American lives. Yet, even among the worst off groups, most of the children of immigrants were exceeding their parents' levels of education, if only because the parents' levels were quite low. *All* of the second generation groups earn more than their native contemporaries of the same "race": controlling for age and gender, we found that Dominicans and South Americans earn more than Puerto Ricans, West Indians more than native blacks, and the Russians and the Chinese are on par with native whites. In terms of educational attainment, Dominicans and South Americans are doing better than Puerto Ricans, West Indians are doing better than native African Americans, Russian Jews are doing better than native whites, and the Chinese are doing better than everyone. While their labor force participation was lower than the staggeringly high levels of their immigrant parents, the proportion of the Chinese and the Russian Jews either going to school full time or participating in the labor force was higher than that of the native whites, and among *every* group it was higher than that of native African Americans or Puerto Ricans. Indeed, with the partial exception of the Dominicans, *every* group was closer to the native whites than to the minority natives. And while there are significant differences among the second generation groups in the number who get involved in criminal activity, even in those groups most likely to have had brushes with the law, the male arrest rate is about that of native whites, well below that of native minorities.

The fact that post-1965 immigrants are overwhelmingly nonwhite was one of the primary reasons for speculating about second generation decline. And we did find a troublingly high number of respondents of African descent report experiencing racial discrimination, particularly by the police. This clearly contributed to their feeling uncomfortable with their status as "Americans" and left many alienated from American institutions and life. Yet, even with regard to this issue our findings are not pessimistic. The second generation group most identified as "black" and most likely to experience such discrimination—West Indians—is also the second generation group most likely to participate in neighborhoods and civic affairs, and to be interested in New York politics. Members vote in numbers comparable to native whites, if somewhat below the very high proportion of native African Americans.

This rapid incorporation into American life is rarely attributable to the second generation's social, cultural, or economic ties to their parents' immigrant communities. The group experiencing the most dramatic upward mobility—the Chinese—is, in fact, the one least likely to retain parents' language. In every second generation group we studied, those who work in predominantly "ethnic" work sites earn less than those who work in mainstream ones. At the same time, today's second generation does not seem overly concerned with shedding those ties or losing their ethnic distinctiveness. Like the 23-year-old Colombian American young woman quoted earlier, the second generation is happy, in Alejandro Portes and Ruben Rumbaut's (2001) phrase, to acculturate "selectively," taking what works the best in their parents' communities as well as incorporating the best of what they see around them among their native peers. Many cheerfully report that they do not feel fully a part of their parents' immigrant communities nor do they see themselves as fully "American"—a term they sometimes use to describe the accentless, un-ethnic whites they know primarily through television. Yet, compared with the "second generations" of the past, today's children of immigrants seem remarkably at ease about being between different worlds. Rarely, if ever, do they see their parents' foreignness as posing a serious problem.

This is not to imply that the incorporation of New York's children of immigrants has been unproblematic. Many have received substandard educations in the city's most problematic public schools. While their labor force participation is high, many are entering jobs with little possibility of advancement in an economy of falling real wages. And yet, there is little about any of these problems that seems distinct to their status as the children of immigrants. They are generally the problems of young working-class New Yorkers, and indeed, in most cases, they are less severe among the children of immigrants than among the members of the native minority groups.

Explaining Second Generation Progress

A decade after beginning our study, we find ourselves explaining good outcomes among the second generation, not second generation decline—even though that generation faces a number of substantial obstacles. Among families with children, those headed by immigrant parents are much less likely to speak English at home (only 19 percent do as opposed to 60 percent for those headed by native parents) and may not even understand English (about a quarter as opposed to only 4 percent for the native parents).[2] Only half the parents in immigrant families are U.S. citizens, giving them far less political influence than native parents.

Most crucially, immigrant parents are less likely to be well educated than native parents: a third lack a high school degree compared with one-fifth of native parents, while only a fifth have college degrees compared with a quarter of the native parents. As a consequence, they have less income. Immigrant parents had a mean household income of $54,404 in 1999 compared with $73,983 for the native parents. Thus young people growing up in immigrant families have parents with less English facility, less education, less political clout, and less income than those growing up in native families. It would be surprising if these factors did not constitute barriers to their progress.

Yet, despite these barriers, the second generation is generally doing better than natives of comparable racial backgrounds. Why is this so? In our book on the second generation, we highlight three reasons for second generation advantages (Kasinitz et. al. 2008).

The first reason is a very important and obvious factor that is nevertheless consistently overlooked. Immigrants are a highly selected group. They are the people who have the drive, ambition, courage, and strength to move from one nation to another.[3] The second generation individuals are, therefore, the children of exceptional parents. Whatever the measurable characteristics of the parents that put the second generation at risk—low education, low skills, low incomes, poor language skills, et cetera—the unmeasurable characteristics of immigrants make them different kinds of parents, mostly in ways that have to be advantageous for the second generation. Thus a Dominican father with a second grade education who does not speak English may appear at first glance to have similar characteristics and fewer resources than even the least well off New York – born Puerto Rican father. Yet, the immigrant parent has other qualities that separate him from all the other uneducated Dominican men who stayed home, and contributed to his success in migrating to New York. His lack of education may not have the same negative effect on his ability to instill a desire for education in his son or daughter. So too, a poorly educated Chinese waiter in New York City is quite different from the many comparable men in China who did not make the journey to New York—he has overcome extraordinary obstacles to change his lot in life. That drive to better his situation is something he is likely to transmit to his children. Thus when comparing natives and second generation individuals, it is important to remember that while these individuals do not constitute a selected population, the parents who raised them surely did.

Second, much of the second generation is uniquely positioned to take advantage of and profit from the civil rights era institutions, affirmative action, and polices promoting diversity that were, ironically, designed to redress injustices suffered by members of native minority groups (see Kasinitz in this volume). Because the children of immigrants are mostly from societies in which they were part of the racial majority (unlike, say, the children who came as refugees from Central Europe in the 1930s and 1940s), the second generation is far less encumbered by the residue of past discriminatory practices. While covert racist practices and assumptions obviously do affect the lives of second generation immigrants—the second, third, or fourth generation Asian American professional "complimented" on his command of English or asked when she is "going home" are the archetypical examples—such practices and assumptions are less pernicious and less pervasive than those confronting minorities whose caste-like subordination has been central to the structure of American society. Ironically, the children of nonwhite immigrants have been better positioned to benefit from the delegitimation of the overt white supremacy and *de jure* racism that the civil rights struggles achieved than were longstanding U.S. minorities—particularly African Americans (see Kasinitz in this volume).

Yet, perhaps the most important advantage for second generation individuals has less to do with their parents or pro-diversity institutional change than with their own unique social position and their ability to develop creative strategies for living their lives. Children of immigrants have often been described as being "torn between two worlds." Social scientists—and immigrant parents—frequently worry that a group navigating between two cultural systems and particularly between two languages may never be completely competent in either. It is often feared that growing up in a world in which your parents have difficulty guiding you into adulthood—having come of age in a different society and culture—can lead to confusion, alienation, and reversal of authority roles within the family. Indeed, in the early twentieth century, many second generation children of European immigrants coped with these competing expectations and world views by rejecting their parents' embarrassingly "foreign ways" and trying to become very "American." Yet, while our respondents occasionally noted the tension between their parents' culture and what they saw as the American world view, rarely did they see this as a problem. Perhaps because of today's ethos of multiculturalism, most of the young people we spoke with believe they can pick and choose which aspects of which cultural model to adopt.

Traditional, "straight line" assimilation theory implies that the children of immigrants will do best when they "assimilate." This may have emotional costs, but in the end, the children of immigrants will come to share the "native advantages" over their immigrant parents (Warner and Srole 1945). Richard Alba and Victor Nee's (2003) contemporary reworking of this notion greatly improves this model, excising its prescriptive aspects, emphasizing the retention of some elements of ethnic culture, and stressing how the U.S. culture is remade in the process. Yet, they too generally see the second generation as coming to share the advantages that come from incorporation into the increasingly multicultural "mainstream." By contrast, one of the central insights of segmented assimilation theory is that resisting Americanization can be helpful for the second generation. This view sees disadvantages for those members of the second generation assimilating into disadvantaged "segments" of the native population, as well as advantages for members of the second generation who, by keeping assimilation at bay (or only partial), can continue to share the "immigrant advantages" of the relatively better-positioned immigrant communities.

Clearly, today's second generation reflects examples of all of these models. However, to these we would add the notion of second generation advantage—advantages that in some situations the second generation individuals have over both their immigrant parents and the native population, advantages tied to their position in between different social systems. This combination allows a creative and selective combination of foreign and American culture that can, at its best, be highly conducive to success—conceptualized here as socioeconomic mobility or exceptional accomplishments or creativity. In developing a strategy or course of action in today's society, it is not a question of whether being foreign or being American is "better." The advantage lies in having a choice—and the knowledge that one does have that choice. Of course, not everyone chooses well, and different groups clearly have different options, depending on both parents' position and the segment of American society into which they are being incorporated. But, other things being equal, seeing choices where others may see mandates and prescriptions is, in itself, a significant advantage. And while puritans of various stripes are generally more comfortable with the coherence of traditional cultural systems, New York, more than most other cities, has generally honored hybridity and rewarded innovation.

In the mid-twentieth century, New York became not only one of the world's leading centers of commerce, but also one of its greatest centers of cultural creativity. There were many reasons for this, not the least of which was American economic ascendancy. However, we do not think it is coincidental that this intensely creative period in American music, art, letters, theater, intellectual life, and criticism coincided with a previous second generation's coming of age. Immigrants and their children played a cultural role far out of proportion to their numbers (Hirschman 2005), and New York, where so much of the second generation was concentrated, became a hothouse for intellectual "scenes" and cultural movements, both mainstream and avant-garde. New York gave the children of immigrants the cosmopolitan space in which to make these innovations. And despite the worries of nativists that New York was becoming a place apart from the rest of the nation, in the end, the second generation repaid America with a new, broader, and (we think) better vision of itself. It was, after all, a 1.5 generation New Yorker, Irving Berlin, who penned "God Bless America." (A Russian-born Jew, he also wrote "I'm Dreaming of a White Christmas.")

It is much too early to say whether something like this is happening today. The world is a different place, the second generation is still quite young, and of course, history does not repeat itself. Yet, the creative mixing of American and immigrant forms and immigrant cultures with each other and with the culture of native minorities is already evident in the music, art, dance, and poetry now being produced in hyperdiverse cities like New York, London, and Los Angeles. The greatest spur to creativity in multicultural cities is, after all, not the continuation of immigrant traditions, nor the headlong rush to become similar to the host society, but rather the innovation that occurs in times and places where many different traditions come together, where no one way of doing things can be taken for granted. And, for all of their problems, there is an undeniable innovative energy to be found in the increasingly diverse working working-class neighborhoods of contemporary New York.

This creativity is not just evident in cultural forms but in the kinds of everyday decisions and behaviors of young people who in effect grow up with a dual frame of reference—their parents' norms and the American norms around them. This means these young people can be, and perhaps must be, creative in their reactions to their environment. Indeed, for many situations, these second generation individuals cannot just blindly apply the received wisdom of their parents—as their parents' wisdom is best suited to a very different society. More than most of us, they know that their parents' ways cannot always be theirs. Nor do they unreflectively take up the ways of an American "mainstream" they are only beginning to know. In a multitude of small decisions about so many aspects of their lives, they must choose—between the ways of their parents, the broader American society, and their often native minority peers or perhaps to create something new and different altogether. And while they do not always choose wisely or well, they cannot help but be aware that they have a choice. And that awareness is a considerable advantage.

We often attribute drive and creativity to the self-selection of immigrants or to ethnicity itself. Yet, the real "second generation advantage" is a "distinctiveness advantage" (Sonnert and Holton in this volume); it comes from being between cultures. The creativity inherent in occupying a position at the crossroads of two groups has been widely recognized in a variety of situations, but we believe it has been insufficiently recognized with respect to the second generation. Sociologist Ron Burt describes the situation of being between two social networks as being in a "structural hole." He notes that "opinion and behavior are more homogeneous within than between groups, so people connected across groups are more familiar with alternative ways of thinking and behaving which give them more options to select from and synthesize. New ideas emerge from selection and synthesis across the structural holes between groups" (Burt 2004: 349–350).

This insight is not new; it goes back to Georg Simmel and even John Stuart Mill (Burt 2004: 350). Yet, if anything, it has become more true for today's second generation than it was for the children of immigrants in the past. The ethos of multiculturalism and the reality of globalization and the unprecedented diversity that characterizes cities like New York multiplies the second generation advantage. Contemporary second generation individuals do not feel undue pressure to reject their parents' languages, beliefs, and behaviors. Nor do they feel the need to cling to them to keep the dangers of assimilation at bay. They grew up in a world in which being different can be "cool," and they insist that they are free to assert certain aspects of their parents' ways and to reject others—thus allowing this cultural creativity to flower.

The creativity of the second generation comes from the ability to meet the needs of the structural environment people find themselves in with a wider repertoire of options for beliefs and behaviors than people who have a single frame of reference—those who grow up in the same society that their parents grew up in. Of course, being between two or more cultural systems can also sometimes be a negative. And the extent of the advantages in combining two sets of norms does depend on which immigrant norms the second generation person draws on and which segment of U.S. society he or she is being incorporated into. Some groups obviously have better options than others. We are not suggesting that the positive side of being between cultural systems outweighs all of the limitations that the most disadvantaged of the second generation face. This "second generation advantage" is one factor among many shaping the young people's lives. Yet, we are convinced that, most of the time, it is on the positive side of the balance sheet, and that many previous observers have either ignored its impact or been too quick to see it as a negative.

This creativity that comes from being between cultural systems could not flower if the local context in which these young people grew up taught them to disparage the culture of their parents. This is where both a temporal change since the last wave of immigration and the particular ethos of New York play a part. For earlier European immigrants there was a strong pressure to Americanize, and this was often felt by young people as a strong disapproval of their parents' cultures—the educator Leonard Covello famously observed that when he was growing up becoming American meant learning to be ashamed of his Italian immigrant parents (quoted in Iorizzo and Mondello 1980: 118).

How different things are today! Few of our respondents were uncomfortable or ashamed. They did not feel torn between two worlds—indeed, they felt often proud of the ways in which they bridged two worlds, of what Monica Boyd and Elizabeth Greico term their "Triumphant Transitions" (1998). This is both because the model of Anglo-conformity that Marcus Lee Hansen (1990) described has been replaced with a strong appreciation of multiculturalism and because of the particular ethos of New York City.

There the second generation inherits an environment where the second generation advantages work to particularly good effect. While these young people feel the sting of disadvantage and discrimination, they move in a world where being from somewhere else has long been the norm. For them being a New Yorker is being both ethnic and American, both different from whites and from their immigrant parents. And in this feeling they are reaping the benefits of New York's long history of absorbing new immigrants and being changed itself by this absorption. As Nathan Glazer and Daniel Patrick Moynihan put it in 1963:

> New York is not Chicago, Detroit or Los Angeles. It is a city in which the dominant racial group has been marked by ethnic variety and all ethnic groups have experienced ethnic diversity. Any one ethnic group can count on seeing its position and power wax and wane and none has become accustomed to long term domination, though each may be influential in a given area or domain. None can find challenges from new groups unexpected or outrageous. The evolving system of inter-group relations permits accommodation, change and the rise of new groups.

This has persisted despite the nonwhite origins of most new immigrant groups. No doubt New York City still has an entrenched white establishment that can trace its roots in the United States back many generations. But the new second generation rarely encounters such people on the job; in the unions; or around the neighborhoods, schools, and subways of New York. Instead, they see a continuum of "whites" who trace their origins to Italy, Ireland, Germany, Russia, Poland, Greece, or Israel. If Italians are yesterday's newcomers and today's establishment, then maybe Colombians are the new Italians and, potentially, tomorrow's establishment. New Yorkers, old and new, are happy to tell themselves this story. It may not be completely true, but the fact that they tell it, and believe it, is significant and may serve to help make it come true.

Looking into the Future: Sustaining Advantages

Will the social mobility and general optimism we found carry over to the third and fourth generations, or will we see stagnation or perhaps a reversal in fortune for the grandchildren and great grandchildren of the immigrants? The latter is the pattern we see with the Puerto Ricans. Academics and policy makers have not paid nearly enough attention to Puerto Ricans in recent decades. Dwarfed by new immigration, Puerto Ricans have often disappeared statistically into the Hispanic category. Neither immigrants or natives, their special political status also allowed them to fall through the cracks—the rediscovery of urban poverty in the United States in the 1980s and 1990s has focused almost exclusively on African Americans, while the new immigration literature has left Puerto Ricans out of the picture entirely. While off stage in New York at least, the Puerto Ricans' situation has deteriorated. The poorest group in our sample, they show distressing evidence of persistent poverty and intergenerational socioeconomic decline. Perhaps this is because the special selectivity of the families of migrants described earlier does not apply to the third generation families our respondents grew up in. Perhaps there is a reverse selectivity effect with the more successful Puerto Ricans out-marrying and losing their identity as Puerto Rican, or moving out to other states where they fall outside our sample. But the New York-based sample we spoke with is doing poorly. And racial discrimination, poor urban schools, language issues, and dysfunctional families all play a part.

Another clear pattern in our findings is deeply troubling. Race and racial discrimination continue to shape the life chances of second generation respondents with dark skin, who can be confused or associated with, or who see themselves as becoming, African Americans. While we do not find much evidence of "second generation decline," the continuing disadvantages faced by native African Americans, the status of the New York – born Puerto Ricans, the poverty and incarceration of many second generation Dominicans, and the high levels of discrimination reported by even the relatively well off West Indians clearly point to the possibility of "third generation decline." Because it encapsulates a complex dynamic of scarce family resources, high obstacles to success, and a risky environment, race still counts very much in New York City. Although many children of immigrant minority parents manage to avoid its worst impacts, that does not lessen the sting on those who cannot.

There is a distinct possibility that some portion of Dominicans' and West Indians' will experience marked downward mobility as they become less distinguishable and as residual immigrant and second generation advantages fade in the third or fourth generation. Our second generation Dominicans are better educated than their parents—but their parents had such low levels of education that it took only a modest increase to show significant progress. Dominicans share with Puerto Ricans a multiracial heritage and are often subject to racial discrimination. And West Indians, despite higher incomes and better educations, are the most likely to report experiencing discrimination from the police and in public places, where their interactions with whites seem little different from those of African Americans.

While we began our study worried about downward mobility of some of the children of immigrants, we now feel that, in some ways, it is the opposite problem that is actually a greater cause for concern. It has become clear that the relative success of the children of immigrants is now overshadowing the extent of continuing poverty and discrimination, limited opportunities, staggering rates of incarceration, and the general social exclusion of large segments of the *native* minority youth population. When elite colleges point with pride to their increasing "diversity" and to the growing numbers of "blacks" and "Latinos" among their students and faculty, it is easy to overlook how much of that diversity is provided by the growing numbers of immigrants and their children, and how little by the descendants of American slaves or by long-present Puerto Ricans or Mexican Americans. And when institutions like the CUNY colleges or New York's selective public magnet schools express concern over their declining "black" and "Latino" enrollments, it is easy to miss how much more dramatic those declines would be, but for the children of West Indians, Dominican, and South American immigrants.

It is my hope that a volume such as this that examines the pathways to success of many second generation individuals identifies strategies and public policies that can be mobilized to aid those children of immigrants who are less successful than most, that help prevent third generation decline, and that ameliorate the position of native minorities.

Notes

* This chapter is based on findings from the jointly authored book, *Inheriting the City: The Children of Immigrants Come of Age*, by Philip Kasinitz, John H. Mollenkopf, Mary C. Waters, and Jennifer Holdaway (Harvard University Press and Russell Sage Foundation Press, 2008).

1. The New York Second Generation Study began in 1999, with telephone interviews conducted with eight different random samples totaling 3,415 men and women aged 18–32 living in New York City, the inner suburbs of New York, and New Jersey. About two-thirds of second generation respondents were born in the United States, mostly in New York City, while one-third were born abroad but arrived in the United States by age 12 and had lived in the country for at least ten years. Concurrently, we interviewed 10 percent of respondents in depth in person about their life histories. Finally we fielded six postdoctoral ethnographers for a year at sites where second generation and native young people were likely to encounter each other (see Kasinitz et al. 2004). The respondents' parents came from five sending regions: the Dominican Republic; Colombia, Ecuador, and Peru (called South Americans in our text); the English-speaking countries of the West Indies, China (including PRC, Taiwan, Hong Kong, and the Chinese diaspora); and the former Soviet Union (Jewish immigrants);. Children with parents having these foreign origins comprised 40 percent of the 1.5 and second generation population aged 18–32 in the defined sample area in 2000. We also included three native comparison groups: native-born whites, native-born blacks, and mainland-born Puerto Ricans. This research was generously funded by the Russell Sage Foundation; the Mellon, Rockefeller, and Ford Foundations; the United Jewish Appeal Federation; and the National Institute of Child Health and Human Development (NIH grants 5RO3HD044598-2 and 999-0173).
2. The data come from the 2000 Census 5 percent PUMS (Public Use Microdata Sample) for New York City and include the individual records for the household head, spouse (if any), and children for households with one or more own or related children.
3. A somewhat modified dynamic applies to refugees.

8

The Changing Role of Race and Ethnicity in the Incorporation of Refugees, Immigrants, and their Children

Philip Kasinitz

Looking at the achievements of today's young immigrants and refugees there is an inevitable temptation to make comparisons with immigrants and refugees of the past, particularly the immigrants of the last great wave, who arrived in the United States roughly between 1882 and 1924, and the refugees who arrived immediately before and after the Second World War. When we do this, we usually compare the immigrants and refugees themselves with their predecessors, and try to explain why *they* are similar or different. Yet equally important is the question of how American society has changed. As Alejandro Portes and Ruben Rumbaut (2001) put it, the "context of reception" is crucial in shaping an immigrant group's success.

The implications of racial and ethnic identities and the roles such identities play in American society, and particularly in higher education, have changed dramatically in the years since the early and mid-twentieth century when large waves of immigrants and refugees were seen. This changed context of reception has significant consequences for young immigrants and refugees, and particularly for the children of immigrants and refugees.

In this chapter I will briefly look at two specific examples: the role of community organizations in the reception of Russian Jewish refugees and the consequences of the "post-*Bakke*" extension of affirmative action policies (broadly conceived) to a variety of groups, as "diversity" replaced "justice" as the main rationale for such policies. Much of the data come from the "Second Generation in Metropolitan New York" study funded by the Russell Sage Foundation—which I codirected with Mary C. Waters, John Mollenkopf, and Jennifer Holdaway. Many of the ideas behind this study originate from our joint endeavors, and much of the credit for any of them that make sense probably should go to my collaborators (see Kasinitz et al. 2004, Kasinitz et al. 2008, see also Waters, in this volume). However, this particular analysis is my own, and blame for its shortcomings should be attributed accordingly.

Russian Jewish Refugees and Community Organizations

Today's immigrants and their children face an American society that is different in many ways from the one that confronted their predecessors in the early to mid-twentieth century. The first difference is that the earlier immigrants and refugees were, in fact, successfully incorporated into American society. By and large they "assimilated"; in many cases they were dramatically upwardly mobile. Yet they also changed and broadened American society in the process (see Alba and Nee 2003).

The fact that the previous immigrants and refugees have been largely successful changes the context for today's newcomers in several ways. First, there is the cultural shift from what Milton Gordon (1964) described as the "Anglo-conformity" model of incorporation dominant in the late nineteenth century to a version of what Lawrence Fuchs (1982) called "civic pluralism" that had become dominant by the late 1960s. This greatly reduced the pressure toward cultural conformity on the part of newcomers. Today, many forms of diversity are not only tolerated, they are celebrated, particularly by institutions of higher education, in ways that would have been hard to imagine in earlier times. Perhaps this cultural shift has not always been beneficial. While revisionist historians have bemoaned the pain and cultural loss experienced by the children of early twentieth-century Jewish immigrants in the assimilationist cauldron of New York's public universities (Gorelick 1981), Nathan Glazer has responded that his CCNY classmates felt little need to have their Jewish experience represented at the institution—and indeed most would have probably found attempts to do so condescending and insulting (Glazer 1997). Still, times change—and the pressures for cultural conformity have been greatly reduced. In the 1940s, showing up on the Harvard or Williams campus wearing a sari, insisting on halal food, removing one's shoes when entering a dwelling, or even speaking with an accent would likely have subjected a young person to considerable social isolation. Today, not only are such things unremarkable, they would, if anything, encourage an invitation to an officially sanctioned campus club made of people who share these ways, paid for by student activities fees.

In addition to the greater acceptance of cultural difference on the part of the broader society is the direct presence of earlier immigrants and refugees in positions of authority in educational institutions. I am not sure what effect this has—it has, to my knowledge, not been studied—but I find it hard to believe this has no effect at all. It would be interesting to know whether immigrants and refugees from earlier times generally feel any particular obligation to mentor newcomers of ethnic backgrounds different from their own. In any case, it does not seem that speaking English with an accent will be a handicap in an academic department in which many senior faculty and perhaps also the chair have foreign accents. And, based on thoroughly unscientific, anecdotal observation, it seems to me it would be hard to find a good physics department at any better than average American university in which this is not the case.

More demonstrable is the fact that the incorporation of earlier waves of immigrants left an institutional infrastructure, some of which is represented in this volume's chapters. This institutional infrastructure includes public institutions, private foundations that serve more or less mainstream constituencies, and groups whose institutional base is within the ethnic communities.

In terms of mainstream institutions, at the elite level, one need only look to the Paul and Daisy Soros Foundation and its fellowships for new Americans. The primary funder of this group is a refugee from an earlier period, although as an organization it is not based within a particular ethnic community, and it serves a wide variety of immigrants and refugees from different backgrounds. These awards, which help to send extremely high-performing students to graduate school, are very generous. The total number of recipients is small, yet the awards are important in that these recipients usually attend high-status schools and may go on to be leaders in their fields. The visible presence of members of new immigrant and refugee groups in such institutions, which the fellowships and others like them facilitate, has effects beyond their numbers. And while the Soros Foundation fellowships are among the very few specifically directed at new Americans, many new Americans benefit from other fellowship opportunities available to "minorities." I will say more about this stretching of the concept of minority by elite higher education institutions and its implications for both the immigrants and the members of native minority groups in the second part of this chapter.

At the more proletarian end of the higher education spectrum is my own City University of New York (CUNY), which has over 200,000 students, half of whom are immigrants (*not* the children of immigrants—who make up a large part of the other half—but immigrants themselves). CUNY's record in educating these immigrants is mixed at best (see Lavin and Hyllegard 1996, Attewell and Lavin 2007). Clearly, CUNY has much to learn about how to encourage immigrant success. But the fact that it exists probably makes a difference. It is also true that CUNY celebrates its immigrants as *immigrants*. Whatever problems they are seen as having, it is clear that "foreignness" is not among them. Indeed, the narrative of immigrant success—an old and a basically optimistic New York story—has become a staple of the university's advertising and of the happy stories we like to tell at graduations and while raising funds. Although it is not always said out loud, this happy story is often implicitly juxtaposed with the university's decidedly less happy experience in educating members of *native* minority groups, although in practice these two populations tend to blur, particularly in the second generation.

In terms of support provided within immigrant communities, New York's Russian-speaking Jewish community is a useful, though probably extreme, case to examine. Since the early 1970s more than 500,000 Jews have left the former Soviet Union for the United States. Approximately 300,000 have settled in the New York metropolitan area. On an average, they are older than other immigrants to the United States, and many have the special needs that come with entering a new society at an advanced age. They are also far better educated than most immigrants: two-thirds of our second generation Russian Jewish respondents in the New York Second Generation Study reported that *both* of their parents have college degrees. However, many arrived without knowledge of English, and in most cases, their professional credentials did not translate easily. Consequently, the parents of many of our respondents began their lives in America working jobs far lower in status than those they had left. The immigrants breathed new life into many Jewish institutions and neighborhoods, but at the same time posed a host of new challenges for the communities in which they settled.

The Jewish community, however, was rich with institutions that were in most cases a heritage of earlier waves of migration. Groups dating from the early twentieth century, such as the UJA-Federation and the Hebrew Immigrant Aid Society (HIAS), or from the efforts to resettle Holocaust era refugees, such as the New York Association for New Americans (NYANA), were well positioned to aid the resettlement process. Indeed, these groups, which had cut their teeth on the problems of immigrant and refugee incorporation, were by the late 1980s suffering real issues of "goal displacement," as social service providers struggled to find new roles in a community that had become quite prosperous. In some cases they turned their attention to other issues, while in others they sought out new populations to serve. But the mission of immigrant resettlement and assistance, particularly around educational issues, was deeply embedded in these groups' original sense of mission, and it was one they took up with gusto when a new group of impoverished Jewish immigrants showed up on their doorsteps in the 1980s. In general, the existence of a large, wealthy, and well-networked "proximal host community" (Waters 1999) played a significant role in easing the transitions of these immigrants.

In many cases the Russian Jewish immigrants, or at least their second generation children to whom we spoke, seemed unaware of the extent of aid their families had received from the organized Jewish community. Of our approximately 300 second and 1.5 generation Russian Jewish respondents, 43 reported that their parents had received assistance from Jewish community groups at some point in the past. This seemed to us astoundingly high. Yet the officials at UJA-Fed and NYANA, who helped pay for the study, scoffed when we reported the figure, contending that they had extended assistance to well over 90 percent of Jewish refugees on arrival. Further, while only 2.7 percent of our respondents reported that they were still receiving such assistance when we interviewed them (at least six years past immigration, but in most cases more than ten years), a far larger number reported using community facilities and programs that are, in fact, paid for by these agencies (many believed that the funding for these services had originated from "the government"). Assistance from these community groups was often mixed with aid from governmental agencies, and helped fill in the gaps in what the government could provide.

The presence of community-based social welfare agencies also played a role in immigrants accessing governmental assistance. While immigrants generally receive less in welfare benefits than they are legally entitled to, in the case of New York's Russian Jews it appears clear that social workers and others from these community-based agencies actually made immigrant families aware of and helped them to apply for benefits. As a result, antipoverty programs became, in effect, resettlement programs. Many Russian Jewish families received some form of pubic assistance, usually supplemental security income (SSI), shortly after arrival. (As refugees they were at that time entitled to benefits other immigrants were not.)

The results, however unintended, were generally good for the recipients. Our Russian Jewish respondents were more likely than any of the four other immigrant groups we studied (South Americans, West Indians, Dominicans, and Chinese) to report having received public assistance at some point while growing up. Their rates were comparable to those of native African Americans and Puerto Ricans. However, they were the least likely group to be receiving such assistance as adults at the time they were interviewed. Welfare, as it turns out, is actually very good for alleviating poverty among people with middle-class background. It permitted the parents of many respondents to spend time out of the labor force while they learned English, received additional training, and generally took steps to reestablish their human capital. It also connected young people with programs that encouraged their attendance in college—although, in fact, with well-educated parents, they usually needed little encouragement. Finally, SSI turned out to be particularly helpful for people in multigeneration families. It subsidized grandparents, who then often became primary caretakers of children, thus permitting both parents to receive training and to reenter the labor force with little childcare costs.

Thus, the presence of a well-organized receiving community turned out to be highly beneficial for the young refugees, in terms of both providing services directly and leveraging its resources by helping newcomers access government resources. Obviously, such advantages are not available for groups who do not have a similarly prosperous "proximal host" community in the United States. This raises an interesting question. Could successful members of previous immigrant and refugee groups be convinced to play such a role for newly arriving young immigrants and refugees who belong to a different ethnic group but whose experience might otherwise be similar? That is, can analogy of experience, rather than ethnicity, become the basis of these sorts of mentoring and sponsorship relationships?

Nonwhite Immigrants and the Institutionalization of Diversity

A second great change in the American context of reception between the time of arrival of the earlier cohorts of immigrants and refugees and the present involves how ideas about race and ethnicity were changed by the African American civil rights movement and its aftermath, and how this change affected societal institutions. This is particularly important for those young immigrants and refugees who, while not African American or members of other long-standing U.S. minorities, are generally classified as "nonwhite" in the United States, a classification that, in earlier times, would have very severely limited their life chances.

However partial their victories and unfulfilled their promise, the movements for racial justice of the 1960s served to delegitimize much of the *de jure* segregation and overt white supremacy that had been central facts of American life and law since the beginning of the Republic. They also contributed a repertoire of ideas and organizational forms for challenging racial subordination. The forms of political action that emerged from the struggle for African American empowerment provided different ways of thinking about racial and ethnic difference. The extension of affirmative action programs to the children of immigrants, the emergence of ethnic studies programs on campus, and the general acceptance for so many official and legal purposes of what David Hollinger (1995) has called the "ethno-racial pentagon" of five "racial groups" in the United States, all point to the extension of an "African American model" to other groups. Ironically, nonwhite immigrants and their children are often better positioned to benefit from this shift than many members of long-standing U.S. minorities, particularly African Americans.

In the second generation study we were struck by how West Indians, to cite the clearest example, have on an average more highly educated parents than do native blacks and therefore are better able to qualify for admission to colleges. Some of West Indian second generation students might not qualify for admission to highly selective colleges without affirmative action in the United States. They have often suffered from growing up in racially segregated neighborhoods and attending public schools that are inferior to those attended by whites, but they were not generally *as* disadvantaged as native African Americans. Since colleges are more than willing to count them as "blacks" in order to demonstrate their own "diversity," they get into these institutions and get the benefits of attending them.

Without the shifts in racial policies that followed the African American led civil rights movement, this would not have been the case. The West Indian immigrants who came to New York in the early twentieth century in large numbers also had higher levels of education than African Americans, on an average (Reid 1939). Chinese immigrants in earlier times also had levels of human capital as high as many who arrive today. Yet in a more overtly racially segregated America, this did not really matter. Their life chances were largely circumscribed by race. Today, the West Indians, Africans, Dominicans, South Americans, and Chinese in our study were all able to take advantage of contemporary institutions' desire for diversity. Their presence in top colleges, corporate workforces, and public office (think President Obama) are often pointed to as evidence of the successes of American multiculturalism and the ability of America to absorb large numbers of immigrants. There is much truth in this assertion, particularly in contrast to Western Europe. But their presence also masks one of the greatest failures of American racial policy—the failure to cope with a heritage of past discrimination, especially with regard to native African Americans. The stagnation and decline of natives in our study—notably of Puerto Ricans and African Americans—stand in sharp contrast to the relative successes of our second generation. Thus, ironically, as an immigrant integration policy, affirmative action generally works well. Yet as a racial justice policy it has often proved insufficient. Why is this so? One need not subscribe to William Julius Wilson's (1977) view that the significance of race in American society is "declining," to share his insight that, for African Americans, racial oppression is multidimensional. It includes overt racist practices, covert racist practices and assumptions within the culture, and residual disadvantage that is the result of *past* racist practices. The civil rights movement greatly reduced, although by no means eliminated, the first dimension. It made less progress against the second and was least successful against the third. The original, individualistic language of "civil rights" was generally not equipped to address the present-day effects of past discrimination. Affirmative action, perhaps the most controversial program to be developed after the civil rights movement, is sometimes billed as a compensatory program to make up for past injustice. But it is also often conceived as a program to guarantee "diversity" and minority representation in education and the workplace. In recent years this diversity rationale for affirmative action has become more common than the focus on redressing past injustice. For elite colleges, in particular, this "diversity" argument was easy to mesh with past practices. After all, their admissions practices had never been solely meritocractic (Karabel 2005). Such institutions had a long history of adjusting admissions procedures to guarantee a diversity of regional origins and the presence of large numbers of the children of alumni, and to ensure the representation of students with a variety of other characteristics. Extending the logic to ethnicity was not a major conceptual stretch. However, as immigration increased after about 1970, it was often not noticed how, in an increasingly diverse America, more and more of the beneficiaries of affirmative action were now immigrants and their children.

Thus, whether or not the children of recent immigrants and refugees are aware of how much the African American struggles against racism have affected their lives, they are well positioned to take advantage of the results of that struggle. Douglas Massey's (2002) study of higher education reveals that immigrants and their children are far overrepresented among the nation's black college students, and this is most true at high-status institutions. Indeed, in many of America's elite colleges, the majority of the black students are immigrants, the children of immigrants, or biracial (see also Massey et al. 2007). Similarly, imprecise definitions have often meant that programs designed for Mexican Americans and Puerto Ricans have been utilized by the children of an ever-broadening category of recent "Latino" immigrants. The replacement of "justice" with "diversity" as a rationale for affirmative action on campus and elsewhere has greatly facilitated this trend. It seems, after all, far less wrenching to admit the children of dark-skinned, but middle-class and often college-educated, immigrants than to truly confront the heritage of America's racial past.

Diversity is, of course, a laudable goal. Elite educational institutions and the nation as a whole are better for having pursued it. However, the emphasis on diversity also allowed these institutions to sidestep the nation's most vexing racial problem—the persistent poverty and exclusion of so many of the descendants of American slaves, who have inherited the scars, handicaps, and cumulative disadvantages of a long history of racial exclusion and discrimination. Children of immigrants have some measurable and no doubt considerable unmeasurable assets and strengths that native minorities do not share (see Halter and Johnson, in this volume). Thus, they are poised to benefit greatly from institutions and programs dating from the civil rights revolution.

In analogy to the case of the Russian Jews, many black immigrants also find themselves in a position to take advantage of resources *within* U.S. minority communities and of what Neckerman et al. (1999) called "minority cultures of upward mobility." This is not entirely new. Although it was scarcely noticed outside the black community, Historically Black Colleges and Universities have long played a role in educating Caribbean immigrants and the much smaller numbers of African immigrants. Some of the best, most prominent members of those communities initially came to the United States to attend such institutions, which also, during the mid-twentieth century, frequently hosted African and Caribbean students and scholars who would eventually play significant roles in their homelands. (Historically Black Colleges also played a small but historically significant role in the incorporation of nonblack refugees by providing employment to several dozen Jewish refugee scholars during the 1930s and 1940s [Museum of Jewish Heritage 2009].) However, until recently, the resources of U.S. minority communities were extremely thin, and there was little reason for any immigrant who had other options to seek them out. This may no longer by true. A case in point is Robert Smith's work on what he terms "Black Mexicans"—young Mexican New Yorkers who seek out both the Afrocentric culture of black advancement and the resources tied to it in their largely African American New York public schools (Smith 2008).

For young immigrants and refugees not regarded as "black," the civil rights model may have even more profound effects. This is often true even for those with little social connection to African Americans. Most of the second generation Asian American professionals studied by Pyong Gap Min and Rose Kim, for example, still report feeling "moderate levels of kinship with African Americans and Latinos" as "these minority communities provide role models in fighting white racism" that many perceive as missing in their own communities (Min and Kim 2002: 177, see also Espiritu 1992, Rodriguez 2002).

Since the 1960s, the heroic model of the civil rights movement was taken up by other groups. Both mainstream civil rights groups and street-fighting Black Panthers inspired Latino, Asian, and Native American equivalents. Black Studies on campus quickly spawned Chicano Studies, Puerto Rican Studies, and Asian American studies—to say nothing of Jewish, Middle Eastern American, and Gay and Lesbian Studies. Affirmative action for African Americans quickly expanded to other protected categories, including women as well as racial minorities (and, at my university, even Italian Americans). At least one major second generation Mexican American writer, Richard Rodriguez, has caused some controversy by making his discomfort with this mode of incorporation a major theme. "As a young man," he writes, "I was more a white liberal than I ever tried to put on black. For all that, I ended up a minority, the beneficiary of affirmative action programs to redress black exclusion." He goes on to note that now, as a writer, "I remain at best ambivalent about those Hispanic anthologies where I end up; about those anthologies where I end up the Hispanic (yet) the fact that my books are published at all is the result of the slaphappy strategy of the northern black Civil Rights movement" (2002: 25–26).

Too often social scientists have simply assumed that being "racialized" as "black" or "Latino" can only have negative consequences for the children of immigrants. This is a view they often share with the immigrant parents. Both the social scientists and the immigrant parents often come to see efforts to retain "old country" ways and ethnic traditions as a way to keep the corrosive effects of American racism and the frightening realities of the American streets at bay. And, of course, they are partially right. Pervasive racism can indeed be soul crushing, and the nihilism of the American ghetto can lead young people down many a self-destructive path. This view, however, oversimplifies what it means to be racialized as members of a minority. African American communities have always been more complicated than this view implies, maintaining their own institutions and paths of upward mobility. In post – civil rights America, the heritage of the struggle for racial justice has given young people new strategies and resources for upward mobility. Thus, for the second generation young people in our study, becoming "black" or "Latino" or "Asian" has positive as well as negative consequences.

I raise the issue of immigrants benefiting from programs and policies designed for, and often by, members of native minority groups with some trepidation. Clearly, some of my brethren on the left will see in this a reason why such programs should be more narrowly restricted to members of native minority groups while many on the right will see in this an argument for the basic wrongheadedness of the affirmative action approach and a reason why such programs should be eliminated altogether. It is probably the case that the perception of affirmative action as a "bait and switch" maneuver—with the benefits going to the children of often elite immigrants rather than to impoverished African and Latino Americans—has eroded the always weak support such programs have had among whites.

Nevertheless, I would argue that the fact that affirmative action and other programs to encourage "diversity" in higher education have worked well for immigrants, refugees, and their children is reason enough that such programs should be supported, even if that was not their original intention. These young people are also hampered by racial discrimination, substandard schools, and a lack of knowledge about the American educational system. Ambitious and often coming from families that invest a great deal in the success of the next generation, these young people are well positioned to make use of such programs. In addition, the government should continue to monitor and fight both overt and subtle racial discrimination in housing, jobs, and schools and by the police. Discrimination is a fact of life for dark-skinned young people, but it feels very different if they know that the law will protect them and that their society does not countenance such behavior.

At the same time, I must conclude these generally optimistic observations with a pessimistic note. Affirmative action and other attempts to encourage diversity in higher education have proved insufficient to address the depths of poverty and exclusion among native minority youths, particularly when they are forced to compete with new immigrants, refugees, and their children. My collaborators and I began our study of the children of immigrants concerned that many were experiencing "downward assimilation" and coming to share the fate of long-standing American minority groups (Gans 1992, Portes and Zhou 1993). Today, however, I have the opposite worry. I am concerned that the relative success of young immigrants, refugees, and their children—often in settings designed to create opportunities for native minority members—is now obscuring the depth of continuing poverty, limited opportunities, and social exclusion of segments of the native minority population.

This is not to argue that facilitating the incorporation of newcomers and addressing the persistent disadvantage of long-standing U.S. minorities are not both worthy goals. They are. They are not, however, the same goal, and they cannot always be achieved using the same methods. It is also not to argue that we must forgo one of these goals in order to achieve the other. At times, efforts to accommodate diversity and to promote racial justice may come into conflict, but they are not fundamentally contradictory. It is important that we do not pretend to be doing one, while actually accomplishing the other. The challenge is to do both.

9

Young, Gifted, and West African: Transnational Migrants Growing Up in America

Marilyn Halter and Violet M. Showers Johnson

Belief in the availability of almost unlimited opportunities in the United States and confidence in the outstanding performance of immigrants are key patterns of thought that have played significant roles in the adaptation of West Africans to life in the United States, and are crucial for understanding and evaluating the success of young ("1.5 generation") West African immigrants and refugees as well as of the second generation. Admittedly, these patterns are not unique to West Africans. However, discrete premigration histories and traditions, the specific trajectories of immigration from that region, and the racialization of the identities of the newcomers in the United States bring some distinctiveness to the West African story. No meaningful examination of this narrative fails to see the influence of an amalgamation of homeland traditions, collective memory, and social capital—a fusion that has mentally translated into a "psyche for success." What is this phenomenon? How was it cultivated? How is it acquired and exhibited by young immigrants and the American born of West African heritage? What are its distinctive variations among particular nationalities? Finally, to what extent is it an effective strategy for success?

Who Are America's West Africans?

The term "West African" connotes a pan-ethnic conglomerate, with color as well as some aspects of regional and historical experience in common.

The nations included under this rubric—Benin, Burkina Faso, Cape Verde, Côte d'Ivoire, Gambia, Ghana, Guinea, Guinea-Bissau, Liberia, Mali, Mauritania, Niger, Nigeria, Saint Helena, Senegal, Sierra Leone, and Togo—represent the legacy of colonial intervention that coerced peoples of a wide variety of traditions, cultures, languages, religions, and worldviews who otherwise would not have necessarily grouped themselves in this way. Nonetheless, since this was the part of Africa where much of the transatlantic slave trade was conducted, there is a continuity between African Americans who are descended from slaves in the United States and the newest arrivals from Africa. Most West African newcomers, having migrated from former British colonies, are English speaking. Smaller numbers hail from Francophone Africa, while those from the Republic of Cape Verde speak Portuguese.

Beginning in the late nineteenth century, initially pulled by the labor needs of the American whaling industry, immigrants from the Cape Verde Islands left their drought-stricken archipelago located off the coast of Senegal, islands that had long been colonized by Portugal, to make southeastern New England their new home. These Afro-Portuguese settlers are particularly noteworthy as they represent the first voluntary mass migration from Africa to the United States. Cape Verdeans are still making the transatlantic journey, but especially over the last three decades they have been accompanied by a broad range of newcomers of diverse ethnic and national origins from the other major West African sending countries of Nigeria, Ghana, Liberia, Sierra Leone, and Senegal, as well as smaller numbers of arrivals hailing from the other countries of that region. Taken together, this movement represents a kaleidoscopic variety of ethnic, linguistic, and religious groups that are transforming the ethno-racial landscape of American society.

Of the nearly 1.4 million foreign born from Africa in the United States today, over a third (36 percent) are from the region of West Africa.[1] Those hailing from Nigeria comprise the largest population of West Africans currently living in the United States, followed by arrivals from Ghana, Liberia, Cape Verde, and Sierra Leone. During the second half of the twentieth century, the flow of legal West African migration shifted course. Previously the prime receiving countries for would-be immigrants had been the former colonial powers of Britain and France; now the United States has become an increasingly popular destination. While immigration policy became steadily more restrictive in Europe, it was liberalized on this side of the Atlantic, especially in the areas of family reunification and the criteria to claim refugee status. A critical factor in redirecting the African diaspora was the prolonged recession in Europe during this period while the United States was experiencing growth in the economy.[2]

Since the 1930s, but especially after 1960 and the overthrow of colonial rule, West Africans have been voluntarily migrating to the United States primarily to obtain a higher education.[3] Many would return to their home countries once their schooling was complete. In recent years, however, the unrelenting cycles of political conflict have pushed many West Africans to relocate on a more permanent basis. Oftentimes, endemic poverty and economic collapse have accompanied the political unrest to create crisis conditions that further impel the migrants to uproot. They come to the United States seeking better economic and educational opportunities as well as to reunify families. Among those migrating as refugees, the earliest group was Nigerians escaping the 1960s Biafran War. However, more recently, it has been Liberians and Sierra Leoneans who have gained refugee status, fleeing escalating levels of political oppression and persecution. Prior to moving to the United States, many new immigrants and refugees had been involved in transborder migrations within West Africa, and in some cases, to other regions of the continent. The journey to the United States is often the last step in an experience of multiple dispersals.

According to the 2000 Census, for the first time in its history, immigrant children are the fastest-growing segment of the U.S. population. West African populations are no exception to this trend. Not surprisingly, given the extent of the war-related deaths, political chaos, and rampant disease resulting from the nearly 15 years of civil conflict in Liberia that only ended in 2003, the foreign born from Liberia, many of whom came with refugee status, count the highest percentage (20 percent) of those 19 years of age or under in their immigrant group.

A "Psyche for Success"

Accomplishments of West Africans as voluntary immigrants to the United States started with the first group from that region, young Cape Verdeans who worked hard in maritime-related jobs, in the textile mills, and on the cranberry bogs and who were industrious enough to establish a permanent settlement on this side of the Atlantic.[4] While this initial influx made an impact, the evolving narrative of West African success is heavily weighted with accounts of the English-speaking student pioneers who came decades after. Without a doubt, the protagonists are men like Kwame Nkrumah, who became the first prime minister of Ghana, and Nnamdi Azikiwe, the first president of Nigeria. Their accounts of their experiences in the United States in the 1920s and 1930s give useful glimpses into the small, little-known African émigré communities of the pre – Civil Rights era. Students like Nkrumah and Azikiwe were eager to go back to participate in the tasks of independence and nation building. But by the 1980s, disillusioned by the political and economic woes in their countries, West African students in the United States, whose numbers had increased considerably, became reluctant to return and many remained, acquiring desirable professional positions such as teachers, engineers, and accountants.[5]

The "student-immigrant" tradition is a central element in the "psyche for success." Adama K., a second generation Sierra Leonean, recalled how her father would stress that "West Africans in this country are among the few groups of immigrants whose communities were laid not by 'uneducated workers seeking work' but by bright people who came to further their education."[6] Quantitative data support this claim: the 2000 Census showed that African immigrants are among the most highly educated groups in the country.[7] Education as the most effective avenue for success is a conviction embraced by the adult immigrant generation and transmitted to young immigrants, and second generations. As the West African immigrant narrative emphasizes, this creed is rooted in their premigration settings. Western education has been valued and has proved to be a formidable criterion for upward mobility since the days of European missionaries. "Education for success can only be achieved through hard work" is an aphorism widely embraced by West Africans in urban areas and even in the villages. As many are quick to point out, although Western education is a European import, hard work is at the core of African traditional life. Thus, it is this blending of a European product with an African cultural resource that shaped the premigration backgrounds of the majority of the adult immigrants and is influencing the lives of their children in America.

Among the Cape Verdean population, however, the "student-immigrant" component of the narrative is less pronounced. Cape Verdeans, past and present, have been pushed and pulled by similar factors. Economic necessity at home and economic opportunity abroad as well as family reunification drive the dynamics of their diaspora. However, the desire to seek a better education plays more of a role today in motivating migration than it did a century ago. Moreover, recent Cape Verdean arrivals are already much more widely educated—most adults enter having completed a high school education in Cape Verde—than those who came in the first wave. In addition, current newcomers are much more likely to be able to speak English since English language classes are a required part of the curriculum in Cape Verdean schools, beginning in ninth grade. Once in the United States, like their Cape Verdean American counterparts, many more are going on to college. Consequently, for the first time in the history of Cape Verdean settlement in the United States, a significant proportion of young adults are receiving higher education.

The Role of "Intellectual Subcultures"

How is the tradition of success through education and other aspects of the "psyche" transmitted to and maintained in the children? The communities formed in the diaspora are the foremost conduits for these processes. Kadiatou K., who fled Sierra Leone with her teenage daughters at the height of the civil war in 1998, gives some useful insights: "What we have created in the United States are 'intellectual subcultures.' Therefore, we cannot hand over our children to America. We must keep them in our subcultures where they can benefit from our traditions and values and be ready for America."[8] Such concerns speak directly to the fear that the immigrant children will fall into the oppositional culture that has developed among some disaffected American-born adolescents, especially among the black and Latino populations, whose attempts to attain upward mobility through educational pursuits have been so thwarted by their dire socioeconomic circumstances that it has led them to reject academic aspirations and to scorn scholastic achievement among their peers.[9]

The majority of young West Africans, both immigrant and second generation, are closely connected with their respective ethnic enclaves through the family and other institutions, including church, mosque, and a variety of associations. Many view their communities as vital institutions of teaching and learning—intellectual subcultures—within which the ethics of work and success are nurtured and transmitted. The family attempts to oversee the children's adaptation, from selecting their circle of friends, insisting on the retention or learning of African languages, to stipulating appropriate social activities. West African immigrants have attempted to replicate traditional family structures like the extended family in their new homes. The children are exposed to the influences of a host of uncles, aunts, grandmothers, and grandfathers. While some of these are biological kin, many of these "relatives" are friends and acquaintances of the parents. A number of the relationships are new ones forged in the United States. Such efforts to reconstruct a cohesive familial network, even when the original family unit might have been disrupted in the migration process, as can often be the case particularly for refugee populations, mirrors settlement patterns that have been well documented for other immigrant groups such as the Vietnamese.[10]

Other institutions, even if they do not blatantly dictate to the children, attempt to steer them toward the values of the homeland cultures. The religious institutions have developed ministries that are specifically devoted to young people, such as the Masjid Lwabahu, a Gambian mosque that started in a small apartment in the Bronx and runs an after school program for children to study the Koran in Arabic. Secular associations also address the advancement of the youth in their mission and goals. For example, the Nigerian Women's Association of Georgia (NWAG) offers scholarships to qualified high school seniors who are Nigerians or of Nigerian descent and organizes an annual summer enrichment program to acquaint the youth and children in the Nigerian community of Atlanta with Nigerian culture, values, and lifestyle. The immigrant press also contributes to the grounding of the young ones. The *African Abroad*, a journal with wide circulation in West African communities in New York, Atlanta, Chicago, and Houston, devotes much attention to youth development. Its most notable contribution is its policy to cover the accomplishments of young West Africans, immigrant and American born, from graduations and awards to service in the U.S. military. The editor, Alex Kabba, believes that not only the young people but entire West African immigrant communities are encouraged and emboldened by such reports.

Identity Matters

For the diverse West African immigrant groups, identities matter. They are crucial in shaping the trajectories of adaptation, assimilation, and socioeconomic mobility. For the majority of the first generation immigrants, transmitting and maintaining seemingly uncomplicated homeland African identities are imperative. They believe that the success, even the lifeline of the community beyond the second generation, depends on the sustenance of viable, diverse African identities. This phenomenon is not new or unique to West African immigrants. Similar patterns among other groups have been identified and examined by scholars such as Mary Waters in her study of West Indians in New York, *Black Identities,* and Milton Vickerman in his work *Crosscurrents: West Indian Immigrants and Race*.[11] Thus, initial trends suggest that West African adaptation patterns resemble those of other foreign-born blacks such as West Indians, whereby the immigrant generation attempts to emphasize distinctive cultural traditions and defies being grouped with African Americans, while second generation individuals are much more likely than their parents to associate and identify with native-born minorities. Alternatively, research on Dominican youth in Providence, Rhode Island, has led one scholar to argue that post-1965 immigrants and especially the second generation are transforming entrenched notions of black/white racial categorization altogether. The young people in the study privileged their ethno-linguistic and cultural background as their racial designation, whether as Dominican, Spanish, or Hispanic, rather than identifying in terms of black or white.[12]

Ruben Hadzide, president of the Ghana National Council of Metropolitan Chicago, echoed the typical first generation concern:

> May I ask you to pause for a moment with me and ponder over what will happen to the next generations of our children if we do not put in place structures that will bring them together when we are no more.... Our identity as a people will be lost and our community as we know it today will be wiped off if we do not lay that foundation! That is the stark reality.[13]

Holding such a perspective has led the immigrant generation to work very hard to erect this formidable foundation. Many communities have devised vehicles to formally transmit the immigrant cultures to the children. By the end of the 1990s, language and culture schools had sprung up in several West African communities in New York, Washington, D.C., Atlanta, and Houston. For example, families banded together to organize dance and cooking classes in their homes in D.C. and Maryland. They explained that they were prompted by the desire to "provide the younger generation with cultural roots that will hold them firmly, help them grow, and give them a sense of identity, which many believe has helped them cope with the difficult transition to life in America."[14] For the overwhelming majority of the West African associations, most of which do not have formal, clearly defined culture schools, the "African socialization" of the younger generation is central to their mission.

What, then, is the "stark reality" for the young members of West African immigrant communities? Contrary to preconceived notions held by the immigrant generation, perceptions and negotiation of identities produce some of the most complex experiences in the youths' multiple worlds. The immigrant subcultures of the parents are not the only context that molds the identities of the 1.5 and second generations. They are shaped simultaneously to varying degrees by the other social spaces that constitute the children's realities—public and private schools, after-school programs, the workplace, restaurants, clubs, theaters and other arenas of entertainment, and, very importantly in the twenty-first century, the variety of cyberspaces. These environments are just as potent, for they are where the young West Africans bump into American obstacles that require a more efficacious set of identities than what a narrowly focused foreign black subculture might provide.

Constructing, maintaining, reconstructing, and making sense of identities have always been important processes in the black American experience. In the unfolding debate over identities in black America, a bifurcation between descendants and immigrants has emerged. According to this division, those who trace their ancestry to the earlier, involuntary diaspora are the descendants; those blacks not born in the United States, and even their American-born children and subsequent generations, all products of the new diaspora, are not truly part of the American black historical experience—therefore, they are not descendants. Thus, who or what, exactly, are the young West Africans in black America? The crucial point that this question raises is that the complexity of their identities is forged by the debates and changing circumstances within their immigrant cultures as well as within black and mainstream America. For example, some students at a predominantly white college, who are American born of Nigerian and Ghanaian heritage, were criticized by African American "descendant" students for joining the African and West Indian Student Association, even though they had been born in the United States and had, according to their chastisers, shamelessly accepted financial aid as American minority students.

The legacy of both colonialism and creolization has continued to divide the West African diaspora over conceptions of race, color, and ethnicity, and thus fundamental identity concerns have been paramount in their life stories. Individual biographies typically reflect these complex diasporic histories. Among West African young people, multiple identifications—whether by continent (as African), by region (as West African), by nationality (as Liberian or Ghanaian or Senegalese), or by the familiar umbrella designations of African-American or black—are variously claimed and expressed. Faced with the complexities of ethno-racial social dynamics, in a variation of the notion of situational ethnicity, individuals sometimes have used several different identity strategies at once in the course of negotiating daily life. The levels of internal diversity can result in intragroup segmentation that challenges the creation of a cohesive community. The deepest divisions are often generational but related to cultural fragmentation as well. Sometimes intergenerational struggles have centered on what is viewed as an erosion of traditional family controls. Related to the intergenerational strife is an increasingly pronounced conflict between the American-born diaspora and the newest immigrants.

As is the case with all Americans negotiating their hyphenated identities, young West Africans find creative ways to blend the variegated cultural components of their fluid ethnic affiliations, especially in the arenas of music, food, fashion, and festive culture. For example, young people attending the celebration of New York's annual Nigerian Independence Day have found creative ways of asserting their Nigerianness while blending both sides of the Nigerian-American hyphen. Dressed in the green and white colors of the Nigerian flag, the slogans on their t-shirts pronounce sentiments like "Born Nigerian, Nigerian Always," "I am 100% Nigerian,"

"Nigerian born in the USA," and "Nigerian Girl Powered by Garri [a staple Nigerian food]."

Conscious Hip-Hop

The values typically associated with the expressive culture of hip-hop would seem like the least likely venue for invoking the psyche of success, and yet so powerful is this mythos in the mind-set of West Africans that this is precisely how one young Nigerian American artist, Madarocka, the "Original" African Queen of Hip-Hop, deftly positions herself. Conscious hip-hop aims to teach those who choose to listen in a format that is considered rooted in African traditions. The hip-hop movement, which began in the early 1970s, was created by the children of African American and West Indian parents who lived through or participated in the Civil Rights movement. Young Jamaican and American-born members of this subculture transformed aspects of black American aesthetics in the process of navigating the complex dynamics of being both West Indian and black in a period of political and social activism.[15] Drawing from African musical traditions, the genre was intended to teach and raise the consciousness of those who listened. This form of hip-hop thrived through the early 1990s, promoting a pan-Africanist stance and delving into topics that affected urban communities of color, such as racism, police brutality and incarceration, drugs (specifically the crack epidemic and the havoc it was wrecking in urban locations), AIDS, black on black violence, apartheid, and teenage pregnancy. While maintaining street credibility has always been an important component of hip-hop culture, rap style took a major turn in the late 1990s when gangsta rap became the dominate force in the global hip-hop world. Mainstream hip-hop, which has become the nation's popular music, employs the rhetoric of the world of "ghetto fabulous." It's all about jewelry, cars, "hot chicks," guns, money, power, sex, and drugs.[16] While the heavy promotion of "ghetto fabulousness" has been seducing the world, another movement of conscious-based hip-hop emphasizing a return to roots and positivity has reemerged to counter the diamonds and glitz lifestyle that some feel has poisoned young people and hip-hop music.

By the end of the twentieth century, children of African immigrants, like those of the Afro-Caribbeans before them, had begun to contribute to this musical form. Inspired by this movement, a trend has emerged whereby Conscious Hip-Hop has become the genre of choice among many young West African rappers. Many of the performers are involved in some capacity with youth, and they are creating music that they hope will address some of the problems that young Africans face in the United States as well as in their native countries. All of the artists are using hip-hop as a means of communicating about issues that are affecting the community at large. Madarocka says that she wants to be "a part of the solution for Nigeria, for D.C. and for the Third World and the slums that resemble them." Drawing from a history of using rap and hip-hop to raise awareness of the issues and problems of the urban world, this new generation of West African artists has come full circle in reviving the aims of the 1970s originators of the hip-hop movement, who sought to utilize their African heritage to draw attention to the issues that faced their communities at home and abroad.

Who is Superior? The Relationship with African Americans

While there is no denying the usefulness of the West African subcultures for the development of their young members, the "psyche for success" is far from flawless. As positively assertive as it is, it can be too lofty, with the tendency to blur realities and elicit resentment and alienation. Sierra Leonean-American Joy Roberts, a high school teacher in Lithonia, Georgia, lamented that its proponents are so preoccupied with savoring the abilities and accomplishments of the young members that they often do not recognize the nature and extent of the challenges that young people face outside the safety of the immigrant enclaves. Roberts, from her vantage point as a member of an immigrant subculture and a teacher in an American public school, has observed the psychological strain felt by teenagers who, believing that they do not fully belong, may try painfully hard to fit in or might, instead, recoil or become belligerent in defiance. Native-born young Americans berate the African born about a number of issues—their appearance and mannerisms, their not being able to speak English or speaking with "funny" accents, and their perceived inferior backgrounds coming from a beleaguered part of the world.

The tension in and outside of school is not simply with American kids; it is more specifically with young African Americans. The tumultuous relationship between the two groups cannot be ignored in any meaningful study of the West African experience in the United States. Almost every American city with an African immigrant presence has seen this tension unfold, sometimes in the most vicious ways. African children have accused African Americans of physically assaulting them, sometimes in the process of robbing them, but very often just because they resent their foreign black counterparts. In November 2005, the situation in Southwest Philadelphia, a predominantly African American neighborhood, came to a head. Tension had been escalating for months before as a result of the rapidly growing Liberian community that had also made its home in that "black" area. A 13-year-old Liberian boy was severely beaten by African American youngsters. The extensive coverage of the incident by the media revealed the extent of the problem and prompted reactions. Members of the Liberian community were sure of the reasons for the assaults, including the attack on the 13-year-old. Sekou Kamara explained: "Some African Americans perceive the growing American-born community as a threat." According to Orabella Richards, a Liberian businesswoman, "There is anger about African immigrants coming here and doing so well." And Varney Kanneh, who claimed that his children had been harassed and attacked in school, believed that "the immigrants make some of the American students look bad."[17]

In turn, African Americans have raised this same point about African immigrants trying to make them look bad. Tampa poet James Tokely admits: "A lot of us do harbor a lot of hostility toward Africans. Many Africans have no idea what our ancestors endured during slavery."[18] Duane W., too, is frustrated by the attitudes of some of the African immigrant children in the metro Atlanta high school he attends: "They act superior and don't care about my sad, but great history."[19] Young West African immigrants and members of the second generation have admitted to displaying this air of superiority, which comes mainly from their exposure to the emerging collective immigrant narrative. In conversations in the safety of their diasporic spaces, including electronic chat rooms, the immigrants ponder and discuss disassociating themselves from "typical African American spaces," including black residential neighborhoods and black institutions of higher learning.

The relationship between young West Africans in America and Historically Black Colleges and Universities (HBCUs) is an important area for study. The experiences of West African immigrants and the second generation at these institutions have been mixed. Some parents discourage their children from considering HBCUs, while others have supported their children's decision to attend those schools. Similarly, among the young West Africans, some relate positive experiences that enhanced their identities as blacks, while others transferred to other institutions because of their dissatisfaction with a "racialized" college experience.

Can contemporary young West Africans in America afford to be simply achievers, relying on their intellectual subcultures? No one is denying the potency of their social capital. Indeed, they are young, gifted, and West African, but so was Amadou Diallo, who was gunned down in 1999 by a unit of the NYPD that was looking for a black man who had sexually assaulted a woman. Most of these young people have no idea of the role of the image of the black male rapist in the history of race and race relations in the United States. Cases like those of Amadou Diallo and Haitian immigrant Abner Louima are evidence that West African and other black immigrants need to understand the African American collective narrative as well.

Sharing their collective histories is one way to improve the relationship between West Africans and African Americans, a consequence that would also facilitate the adaptation of the African-born youths and second generation immersed in the sanctuary of their parents' subcultures. This is one of the main objectives of the Umoja Media Project, a part of the Harlem Children's Zone, a nonprofit agency. Umoja, launched in 2001, was partly a response to the tension generated by the 9/11 tragedy between African Americans of Harlem and African immigrants, especially Muslims. Under this project, young African Americans and the children of African and Caribbean immigrants were provided with the necessary equipment to produce a documentary about their communities and their perceptions of and relationships with others. The project gave the participants a chance to discuss issues of identity and stereotypes. Young West Africans, who have the opportunity to become involved in projects such as this one, could prove to be useful conduits between their African American milieu and their ethnic subcultures.

Making a Difference

Dialogue and sharing collective narratives were also what the women of the NWAG had in mind when they introduced a program to visit schools in the Atlanta area. Through talks, slide shows, performances, and other activities, they aim to "educate students about the culture and heritage of Nigeria and Africa." Similarly, the Cape Verdean Alumni Network (CVAN) regularly sponsors events in conjunction with the consortium of Cape Verdean Student Associations to foster the link between cultural literacy and educational attainment. One such CVAN initiative cosponsored by the Cape Verdean Consulate and the Embassy of the Republic of Cape Verde was a youth workshop with a transnational perspective and the theme of "Culture, Integration, Education and Career," held at Roxbury Community College in Boston where the keynote speaker was the prime minister of Cape Verde. In Providence, where a large population of Liberians resides, the Liberian Association of Rhode Island has been actively engaging the younger members of the community, culminating in an annual Youth Summit. In 2006, the focus was on Youth Confronting Violence, and the association drew on the resources of educators, the government, and social service providers to create a forum for "Real Talk" on this urgent subject.

The initiative for dialogue and understanding is not only coming from the adults. The young people themselves are beginning to address the realities of their "in-between world." Youth for Sierra Leone offers a good example of this trend. Founded by mostly young immigrants and refugees with a few second generation members, this organization, with branches in Washington, D.C.; Atlanta; and Freetown, Sierra Leone, was created primarily to "improve the conditions of young people" in that West African country. Initially, Sierra Leone was its exclusive focus. Recently, however, a discussion has begun about why and how to talk about issues affecting the members in the United States.

It is important that young West Africans involve their families and the other institutions of the ethnic enclaves in whatever progress they are making toward examining and understanding their American experience in its entirety. This does not mean that they should abandon the psyche of success constructed in the immigrant community. This social capital pays dividends but must be used discriminately. They should tap into their subcultures in meaningful ways that not only show their pride and confidence in their West African immigrant values but also their readiness to meet the challenges of their existence as black people in America.

The best of what West African young people can bring with them is to enhance their own odds of making a successful adjustment to the United States while also making vital contributions to American society and culture, just as millions of the young, gifted, and foreign born have done for generations before them.

Notes

1. U.S. Census Bureau (2006) American Community Survey and Census (2000) data.
2. April Gordon, "The New Diaspora—African Immigration to the United States," *Journal of Third World Studies* 15, 1 (1998): 79–103; Arun Peter Lobo, "Unintended Consequences: Liberalized U.S. Immigration Law and the African Brain Drain," in Konadu-Agyemang, Kwado, Baffour Takyi and John Arthur (eds), *The New African Diaspora in North America: Trends, Community Building, and Adaptation* (Lanham, MD.: Lexington Books, 2006): 189–208. As is the case with other populations of the foreign born living in the United States, the figures do not include those who are unauthorized; neither do the population trends account for those arrivals from Africa who hold temporary status (such as student visas) and who extend their legal stay.
3. We acknowledge that the term "voluntary" can be problematic. Increasingly, scholars from various disciplines are insisting that the movement of various African groups as a result of war, political instability, natural disaster, and extreme poverty should be considered as forced migration.
4. Marilyn Halter, *Between Race and Ethnicity: Cape Verdean American Immigrants, 1860–1965* (Urbana, Ill.: University of Illinois Press, 1993).

5. Kwame Nkurmah, *Ghana: The Autobiography of Kwame Nkrumah* (New York: Thomas Nelson & Sons, 1957) and Nnamdi Azikiwe, *My Odyssey* (London: C. Hurst and Co., 1971). The legacy of the notion of student immigrant success among those of African descent actually dates back to the 1920s when Caribbean immigrants were arriving in the United States in significant numbers for the first time. The central narrative among this population was a variant of this mythos as it emphasized an activist immigrant tradition. See Winston James, *Holding Aloft the Banner of Ethiopia: Caribbean Radicalism in Early Twentieth Century America* (New York: Verso Press, 1999) and Violet Showers Johnson, *The Other Black Bostonians: West Indians in Boston, 1900–1950* (Bloomington: Indiana University Press, 2006).
6. Interview with Adama K. by Violet Johnson, September 14, 2006, Atlanta, GA.
7. John Logan and Glenn Deane, "Black Diversity in Metropolitan America," Lewis Mumford Center for Comparative Urban and Regional Research, University at Albany, 2003 http://mumford1.dyndns.org/cen2000/report.html; "News and Views: African Immigrants in the United States Are the Nation's Most Highly Educated Group," *The Journal of Blacks In Higher Education* 26, January 31, 2000: 60; Darryl Fears, "Disparity Marks Black Ethnic Groups, Report Says African Americans Trail Immigrants in Income, Education," *Washington Post*, March 13, 2003.
8. Interview with Kadiatou K. by Violet Johnson, October 3, 2006, Clarkston, GA.
9. See Prudence Carter, *Keepin' It Real: School Success Beyond Black and White* (New York: Oxford University Press, 2005); Xue Lan Rong and Frank Brown, "The Effects of Immigrant Generation and Ethnicity on Educational Attainment among Young African and Caribbean Blacks in the United States," *Harvard Educational Review* 71 (3), 2001: 536–565; Marcelo Suárez-Orozco, "Immigrant Adaptation to Schooling: A Hispanic Case," and John Ogbu, "Low School Performance as an Adaptation: The Case of Blacks in Stockton, California," in M. Gibson and J. Ogbu (eds), *Minority Status and Schooling: A Comparative Study of Immigrant and Involuntary Minorities* (New York: Garland, 1991): 37–61 and 249–286.
10. Nazli Kibria, *Family Tightrope: The Changing Lives of Vietnamese Americans* (Princeton, NJ: Princeton University Press, 1993); Min Zhou and Carl Bankston, *Growing Up American: The Adaptation of Vietnamese Adolescents in the United States* (New York: Russell Sage Foundation, 1998).
11. Mary C. Waters, *Black Identities: West Indian Immigrant Dreams and American Realities* (New York: Russell Sage, 1999), chapter 8 is particularly useful; Milton Vickerman, *Crosscurrents: West Indian Immigrants and Race* (New York: Oxford University Press, 1999).
12. Benjamin Bailey, "Dominican-American Ethnic/Racial Identities and United States Social Categories," *International Migration Review* 35, 3 (2001): 677–708.
13. Ruben Hadzide, address delivered at the banquet to celebrate the fiftieth independence anniversary of the Republic of Ghana, Hyatt Regency, Chicago, March 10, 2007, http://ghanaweb.com

14. Remi Aluko and Diana Sherblom, "Passing Culture on to the Next Generation: African Immigrant Language and Culture Schools in Washington, D.C.," *Festival of American Folklife Program Book* (Washington, D.C.: African Immigrant Folklife Project, 1997).
15. For the Caribbean origins of hip hop, see Wayne Marshall, "Hearing Hip-Hop's Jamaican Accent," *Newsletter of the Institute for Studies in American Music* XXXIV, (2 spring), 2005.
16. Nelson George, *Hip Hop America* (New York, NY: Penguin, 1998): 121–127.
17. Robert Moran, Gaiutra Bahadur and Susan Snyder, "Residents Say beating Fits Widespread Pattern, "*The Phildelphia Inquirer*, November 3, 2005; Elmer Smith, " 'Normal' at Tilden Middle School is different, *The Philadelphia Inquirer*, November 4, 2005.
18. Tracie Reddick, "African vs. African-American: A Shared Complexion Does Not Guarantee Racial Solidarity," *Tampa Tribune*, May 8, 2003.
19. Interview with Duane T. by Violet Johnson, November 15, 2006, Decatur, GA.

10

Mexican Immigrant Legalization and Naturalization and Children's Economic Well-Being

Frank D. Bean, Susan K. Brown, Mark A. Leach, and James Bachmeier

Over the past four decades, Mexican immigration has garnered much of the public and legislative attention devoted to reforming immigration policy in the United States (Bean and Lowell 2004). Part of the reason is that Mexicans have constituted the largest of the country's recent legal immigrant groups. In 2005, for example, 161,445 Mexicans gained legal permanent residency, or 14.4 percent of the all such persons (Office of Immigration Statistics 2006). But much of the focus falls on Mexicans because they comprise such an overwhelming component of unauthorized migration flows. Roughly 300,000 (net) unauthorized Mexicans established *de facto* U.S. residency in 2005, bringing the total number of unauthorized Mexicans to 6.2 million, or 56 percent of all unauthorized persons in the country (Passel 2006). These numbers dwarf those from any other nation. Moreover, many observers have long argued that policies to curtail or "regularize" unauthorized migration should be adopted *before* changes in legal immigration policy are considered, thus ensuring that unauthorized Mexican migration occupies a prominent place in any debate over immigration policy (U.S. Commission on Immigration Reform 1994).

But another issue also boosts Mexican migration to the top of the immigration policy agenda, namely, doubts about the success of Mexicans' economic incorporation. Almost all Mexican migrants arrive with very little money or education and are consigned to the bottom tier of the workforce. Consequently, analysts often conclude that their prospects for joining the American mainstream are dim (Camarota 2001, Hanson 2003). In addition, it has frequently been asserted that the immigrants' children and even their grandchildren and great-grandchildren will remain similarly disadvantaged, as Bean, Brown, and Rumbaut noted (2006). But both immigrants and their descendents change, across time as the former stay longer in the country, and across generations, as they give way to their children and their children's children. Reaching adequate conclusions about Mexican incorporation thus depends not only on assessing what happens to immigrants after they arrive in the United States, but also on studying what happens to the second generation. In this chapter, we introduce new data to focus on how changes in the unauthorized and citizenship status of Mexican immigrant parents relate to their children's socioeconomic status (their acquisition of human capital, occupation, and earnings). Such trajectories not only help to reveal the rapidity with which Mexican immigrants are joining the American economic mainstream, but they also provide policy-relevant information about how pathways to legalization and citizenship relate to economic progress among the children of immigrants.

Our assessment is based on new data from a project that focuses on the children of immigrants in metropolitan Los Angeles. Not only is Greater Los Angeles important for its size—more than 17 million people as of 2004—it is one of the two major immigrant gateway metropolises in the country, along with New York (Sabagh and Bozorgmehr 2003). Nearly a third of LA's population is foreign born, with nearly two-thirds of this group from Latin America (U.S. Bureau of the Census 2006). Greater Los Angeles, more so than any other city, has been a receiving center for Mexicans for generations (Grebler, Moore and Guzman 1970). It is now home to nearly six million persons of Mexican origin, or more than one-third of its population. Most important, it has long been the major urban destination for *unauthorized* Mexican entrants (Bean, Passel and Edmonston 1990). Consequently, California was the state in which the most people legalized their migration status when given the opportunity under the 1986 Immigration Reform and Control Act (IRCA) (González Baker 1997). Los Angeles thus provides a unique metropolitan context in which to examine how changes in the legal and citizenship status of Mexican migrants affect their children.

Given these considerations, what would we expect overall integration processes to look like among Mexican immigrants and their descendants? In general, while unauthorized entry may handicap Mexican entrants, the important changes that occur across subsequent generations may nonetheless eventually make up for the fact that Mexican immigrants often have to begin their socioeconomic climb "from the basement, not the first floor," as Bean and Stevens (2003) put the matter. The disadvantage of starting as an unauthorized entrant stands in sharp contrast to what has often been the starting point for other immigrant groups. The theoretical expectation following from this is that evidence of assimilation may not strongly emerge for the Mexican-origin group until the third or later generations because it is not until then that the unauthorized beginning status of many Mexican first generation persons can be overcome. We do not assess overall incorporation here, but rather introduce data to examine how changes in the unauthorized and citizenship status of Mexican immigrant parents relate to their children's economic status (their acquisition of human capital, occupation, and earnings) and thus to the subsequent course of Mexican economic incorporation.

Why Might Legalization and Naturalization Foster Children's Success?

The Benefits of Legalization

Becoming a legal permanent resident with a "green card" offers obvious tangible benefits to immigrants: legal employment, access to a wider range of jobs, legal protections, financial services, and travel opportunities. Legalization also works indirectly to provide the sorts of stable working conditions and job experience that enhance wages and reduce the necessity for workers to rely only on social contacts for jobs (Massey 1987, Aguilera and Massey 2003). Moreover, permanent residents become eligible to naturalize and thereby gain even greater access to certain kinds of employment and public assistance (Bean and Stevens 2003). Legal status thus constitutes an extremely important milestone in the process of immigrant incorporation, and we hypothesize that, by extension, it also matters for the well-being of the immigrants' children.

The Benefits of Naturalization

In the United States, the naturalization generally requires that a migrant be an adult, a legal permanent resident, and a resident of the United States for at least five years. Immigrants must also demonstrate the ability to speak, read, and write English; pass a test on U.S. government and history; and show good moral character (e.g., not have a felony conviction), all characteristics valuable in the labor market. Those who naturalize also are more invested in the U.S. economy as home or business owners and less likely to emigrate, often because their countries of origin are distant or poor (Bernard 1936, Beijbom 1971, Barkan and Khokhlov 1980, Jasso and Rosenzweig 1986, Yang 1994). Family status is important: those with children are more likely to naturalize (Liang 1994, Yang 1994). Naturalization also varies with age at immigration (Yang [1994] finds a convex curvilinear relationship) and gender (women are more likely to naturalize: Jasso and Rosenzweig [1986], Yang [1994]).

The benefits of naturalization are multifaceted and illustrated by two complementary views about the foundations of citizenship. The first perspective sees citizenship as involving distinctly political-economic rights (Ong 1999), such as voting and access to certain employment and labor market opportunities (Aleinikoff 2001). Those who become citizens can expect to be able to vote and to pursue new job possibilities; in turn, they are assumed to embrace largely uniform national identifications (Schuck 1998, Aleinikoff 2003). This framework on citizenship envisions immigrants individually and quite explicitly naturalizing for political and economic benefits.

A second perspective emphasizes additional bases for citizenship (Feldblum 2000, Bloemraad 2006), some of which may involve the operation of post-nationalist and transnationalist forces, implying in some instances a diminishing relevance of national citizenship altogether (e.g., Carens 1987, Bauböck 1994, Soysal 1994, Jacobson 1996). Scholars in this school of thought have emphasized that citizenship strengthens the immigrants' social and symbolic integration in various ways (Basch, Glick Schiller and Szanton Blanc 1994, Liang 1994, Ong 1999, Portes, Guarnizo and Landolt 1999, Morawska 2001, 2003, Gilbertson and Singer 2003).

Possible Differences in the Effects of Mother's and Father's Status

Mother's versus father's status could have different implications for children's outcomes. Because the migration process is gendered in important ways (Hondagneu-Sotelo 1994, Harzig 2006, Suárez-Orozco and Qin 2006), particularly in the case of Mexican labor migration, mothers and fathers may have different reasons for migrating and different opportunities for legalizing their status. Solo male migrants who legalize may petition for spouses to come from Mexico, with the result that some wives may enter the United States legally even though their husbands were initially unauthorized. Also, wives who enter without authorization may have more trouble gathering the paperwork necessary for legalization (González Baker 1997). Such considerations suggest father's legal status and citizenship may have a greater effect than the mother's status on the acquisition of human capital by the children of immigrants.

Alternatively, mother's status may be more important to children's outcomes because mothers tend to be more involved in the socialization of children (Ortiz and Cooney 1985, Matthews 1987, González-López 2003). Thus, we have no theoretical basis for clearly predicting which of these influences might predominate and treat the impact of parental gender as an empirical question.

Data and Approach

Our data come from a survey called Immigration and Intergenerational Mobility in Metropolitan Los Angeles (IIMMLA), supported by a grant from the Russell Sage Foundation and conducted in 2004. The data collection effort targeted the young-adult children of immigrants from large immigrant groups in Los Angeles and obtained information from 4,780 persons ages 20–40 who had at least one immigrant parent (all countries of origin were included). Because of the centrality of the Mexican origin group to the immigrant experience in Los Angeles, the Mexican sample was designed to be a random probability sample of all Mexican-origin persons (whatever their generational status) residing in households with telephones in the greater five-county metropolitan region. The survey obtained information on parents' migration status, at the time of both entry into the United States and the IIMMLA interview. We also collected data on whether the parents had naturalized. The sample size for the group with at least one Mexican-born parent is 935.

Because we define second generation respondents as those having at least one immigrant parent, it is important to note that the generational statuses of the parents may differ. In a few cases, mothers were foreign born but fathers native born or vice versa, meaning that one parent could *not* have had either a legalization or naturalization experience. Respondents with one native-born parent thus constitute an especially meaningful comparison group, because their parents' cross-nativity marriage may give them a substantial "jump-start" in incorporation, compared with those having two foreign-born parents. We include them in a separate category to provide a useful benchmark for children's economic attainment. As a consequence, we examine six nativity/migration status/naturalization trajectories for the mothers and fathers of the IIMMLA 1.5 and second generation respondents of Mexican origin. These trajectories are (together with the terms we use for them): (1) Native Born: father (mother) is native born; (2) Authorized/Citizen: father (mother) is authorized at entry, later naturalized; (3) Authorized/Green Card: father (mother) authorized at entry, not naturalized by time of interview; (4) Unauthorized/Citizen: father (mother) unauthorized at entry, naturalized by interview; (5) Unauthorized/Green Card: father (mother) unauthorized at entry, obtained legal permanent residency, but not naturalized at interview; and (6) Unauthorized/Unauthorized: father (mother) unauthorized at both entry and interview.

Findings

How have the pathways to legalization and citizenship among unauthorized immigrants affected their children's chances of joining the American economic mainstream? We first should note that calculating the fraction of our respondents whose parents came as unauthorized entrants depends on the number of parents who in fact were immigrants. Roughly 10 percent of the fathers and the mothers were born in the United States and thus could *not* be immigrants, although their children qualify as 1.5 or second generation persons because of the immigrant status of their other parent. In addition, another 119 fathers and 81 mothers never migrated to the United States, and this group constitutes 12.7 percent of the fathers and 8.7 percent of the mothers in the sample (see table 10.1). We omit both of these groups in calculating fractions of 1.5 and second generation persons with unauthorized fathers and mothers. Because the migration status of 69 parents is unknown, we calculate three sets of percentages. One assumes that these parents were unauthorized, and the second assumes they were not. The third percentage excludes this group of parents altogether. The first set of percentages will yield a maximum estimate of the percentage unauthorized, the second a minimum estimate, and the third an in-between estimate.

The resulting sets of three percentages are shown in the first six rows of table 10.2 and reveal that almost half of the fathers of the 1.5 and second generation respondents came to the United States as unauthorized migrants (about 46 percent in the case of the middle estimate). Among the mothers, more than two of every five came as unauthorized migrants, not quite so high a ratio as for the fathers. These estimates are reasonably close to previous ones for the fractions of unauthorized entrants from Mexico eventually settling in California during the 1950s, 1960s, and 1970s (Bean, Passel and Edmonston 1990).

By the time their children participated in the IIMMLA interview, only about 5 to 14 percent of the fathers remained unauthorized. Among mothers, 5 to 6 percent remained unauthorized. If we work with the middle

Table 10.1 Entry Status and Legal and Naturalization Trajectory, Fathers and Mothers of 1.5 and Second Generation Mexican-Origin Respondents

Status at Entry/ Status at Interview	\multicolumn{2}{c}{All}		\multicolumn{2}{c}{Those with Foreign-born Fathers}		\multicolumn{8}{c}{Distribution by Father's Status}						
					Those with Foreign-born Fathers and Known Migration Status		Those with Foreign-born Fathers Who Could Have Migrated or Did to the United States		Those with Foreign-born Fathers and Known to Have Migrated to U.S.	
	N	%	N	%	N	%	N	%	N	%
Status Unknown[a]	60	6.4	60	7.1	–	–	60	8.3	–	–
Never Lived in the United States	119	12.7	119	14.1	119	15.2	–	–	–	–
Not Foreign Born	93	9.9	–	–	–	–	–	–	–	–
Authorized / Naturalized	239	25.6	239	28.4	239	30.6	239	33.1	239	36
Authorized / Green Card	118	12.6	118	14	118	15.1	118	16.3	118	17.8
Unauthorized / Naturalized	152	16.3	152	18.1	152	19.4	152	21	152	22.9
Unauthorized / Green Card	114	12.2	114	13.5	114	14.6	114	15.8	114	17.2
Unauthorized / Unauthorized	40	4.3	40	4.8	40	5.1	40	5.5	40	6
Total for Fathers	935	100	842	100	782	100	723	100	663	100

Table 10.1 (Continued)

Status at Entry/ Status at Interview	All		Distribution by Mother's Status							
			Those with Foreign-born Mothers		Those with Foreign-born Mothers and Known Migration Status		Those with Foreign-born Mothers Who Could Have Migrated or Did to the United States		Those with Foreign-born Mothers and Known to Have Migrated to U.S.	
	N	%	N	%	N	%	N	%	N	%
Status Unknown[a]	9	1	9	1.1	–	–	9	1.2	–	–
Never Lived in the United States	81	8.7	81	9.7	81	9.8	–	–	–	–
Not Foreign Born	98	10.5	–	–	–	–	–	–	–	–
Authorized / Naturalized	300	32.1	300	35.8	300	36.2	300	39.7	300	40.2
Authorized / Green Card	128	13.7	128	15.3	128	15.5	128	16.9	128	17.1
Unauthorized / Naturalized	138	14.8	138	16.5	138	16.7	138	18.3	138	18.5
Unauthorized / Green Card	142	15.2	142	17	142	17.1	142	18.8	142	19
Unauthorized / Unauthorized	39	4.2	39	4.7	39	4.7	39	5.2	39	5.2
Total for Mothers	935	100	837	100	828	100	756	100	747	100

[a] Did not know parent or parent's status

Table 10.2 Percent of Fathers and Mothers with Various Legal and Naturalization Statuses, 1.5 and Second Generation Mexican-Origin Respondents

	Fathers	Mothers
	%	%
Percent Entering Unauthorized	46.2[a]	42.7[a]
	42.3[b]	42.2[b]
	50.6[c]	43.4[c]
Percent Unauthorized at Interview	6.0[a]	5.2[a]
	5.5[b]	5.2[b]
	13.8[c]	6.3[c]
Percent Legalizing of Entrants with Known Status	94.0	94.8
Percent Naturalizing of Known Legal Entrants	66.9	70.1
Percent Naturalizing of Known Unauthorized Entrants	49.7	43.3
Percent Naturalizing of All Known Eligible	62.8	61.9

[a] Only for parents with known entry status.
[b] Assumes those parents with unknown status were all authorized.
[c] Assumes those parents with unknown status were all unauthorized.

estimate for the percentage that legalized, this would mean that nearly nine of every ten parents who were unauthorized entrants had attained legal status by 2004. Overall, it would mean that about 19 of every 20 of the parents who were *known* to be entrants were either legal or had attained legal permanent resident status by the time of the IIMMLA interview. This very high percentage of legal fathers and mothers among the children of Mexican immigrants in Los Angeles provides strong testimony to the successful implementation of the legalization provisions (González Baker 1990) provided by the 1986 Immigration Reform and Control Act (IRCA), and to the overall effectiveness of the legislation's legalization provisions (Bean, Vernez and Keely 1989). While we do not have an exact date for parents' legalization, most legalizations took place in the late 1980s and early 1990s.

How does legal status, including the IRCA legalization that would have been utilized by many of these fathers and mothers, relate to the human capital attainments of their young adult children?[1] In table 10.3, we see that those respondents whose fathers became legal permanent residents (LPR) were *less* likely to drop out of high school (16.9 percent vs. 22.5 percent) and *more* likely to graduate from college than were those whose fathers did not become LPRs. Similarly, they are nearly 13 percent more likely to prefer speaking English at home, more likely to work in jobs with higher occupational prestige (about 7 percent higher on average), and more likely to work

Table 10.3 Indicators of Human Capital and Economic Attainment by Father's and Mother's Entry Status and Legal Status at Time of Interview, 1.5 and Second Generation Mexican-Origin Respondents

	Fathers	Mothers
Respondent Education		
% Less than High School Diploma		
Not Foreign Born	11.8	15.3
Authorized at Entry	13.2	14.0
Unauthorized / Authorized	16.9	15.0
Unauthorized / Unauthorized	22.5	35.9
Status Unknown	26.7	11.1
Never Lived in the United States	37.0	49.4
% Bachelor's Degree or Higher		
Not Foreign Born	19.4	8.2
Authorized at Entry	16.2	19.2
Unauthorized / Authorized	17.3	15.7
Unauthorized / Unauthorized	10.0	0.0
Status Unknown	8.3	0.0
Never Lived in the United States	8.4	8.6
Average Years of Education		
Not Foreign Born	13.5	12.7
Authorized at Entry	13.2	13.4
Unauthorized / Authorized	13.2	13.2
Unauthorized / Unauthorized	13.0	11.4
Status Unknown	12.4	13.0
Never Lived in the United States	11.8	10.9
% Prefer to Speak English at Home		
Not Foreign Born	71.0	80.6
Authorized at Entry	65.5	62.9
Unauthorized / Authorized	50.8	46.1
Unauthorized / Unauthorized	45.0	28.2
Status Unknown	41.7	66.7
Never Lived in the United States	32.8	28.4
Average Occupational Socioeconomic Prestige		
Not Foreign Born	40.4	42.4
Authorized at Entry	42.3	41.9
Unauthorized / Authorized	41.3	41.4
Unauthorized / Unauthorized	38.5	34.0
Status Unknown	39.6	42.6
Never Lived in the United States	38.8	36.9

Average Personal Income
Not Foreign Born	23,194	25,847
Authorized at Entry	23,847	23,466
Unauthorized / Authorized	22,105	20,014
Unauthorized / Unauthorized	16,988	14,218
Status Unknown	19,567	16,056
Never Lived in the United States	17,395	19,685

in jobs with higher earnings (about 30 percent higher than those whose fathers did not legalize). Thus, in general, having a father who became a legal permanent resident (either through IRCA or regular channels) is related to appreciable socioeconomic benefits for the 1.5 and second generation children of the male Mexican immigrants who entered the country initially in an unauthorized status.

The results for mothers are generally similar. Notably, when the mother remains unauthorized, children acquire less human capital than when the father remains unauthorized. More than twice as many of those with mothers who remained unauthorized never received a high school diploma (nearly 36 percent vs. about 15 percent of those who legalized), and none received a college degree. The occupational prestige of respondents' jobs is about one-eighth lower when their mothers remained unauthorized than when their fathers did, and their income is more than $2,500 lower. Only 28.2 percent of respondents whose mothers remained unauthorized prefer to speak English at home, compared with 45.0 percent of those whose fathers remained unauthorized. Such findings tend to support the traditional gender role socialization perspective, that the offspring of mothers who do not become legal permanent residents are more likely to inherit the disadvantages of mother's status and that mother's status may have even stronger effects than father's possibly because many of the mothers who remained unauthorized may have been unmarried when the respondents were growing up.

Are still further advantages related to naturalizing, among both those whose fathers and mothers entered legally and those whose fathers and mothers were unauthorized entrants and then became LPRs? Of the former group, more than two-thirds (66.9 percent of fathers and 70.1 percent of mothers) had naturalized by the time of the interview (table 10.2). Of the parents known to be unauthorized entrants, about half of the fathers (49.7 percent) and slightly less than half of the mothers (43.3 percent) had naturalized. Thus, 20–35 years after most of the respondents' fathers and mothers had come to the country, about three-fifths of the mothers and fathers had become citizens, including many who started out as unauthorized entrants. Again, it is worth noting that most of them would have qualified for LPR status and citizenship by virtue of the legalization programs of IRCA, which created two major pathways to legalization for unauthorized migrants in the country at that time (Bean, Vernez and Keely 1989). Most of the parents of our respondents migrated to the United States during an era when almost all of them would have been eligible for one or the other of these programs. Although we did not obtain data on whether our respondents' parents in fact became LPRs through IRCA's programs, about three-fourths of the unauthorized Mexican immigrants estimated to be in the country during the 1980s legalized as a result of IRCA (Bean, Passel and Edmonston 1990; Massey, Durand and Malone 2002), and many of the IIMMLA parents would have been included in this group.

The substantial degree of naturalization among the IIMMLA respondents' parents also provides an instructive example about the degree to which the availability of pathways to legal status and citizenship matter for the economic well-being of the children of unauthorized immigrants. The human capital and labor market outcomes among the children of immigrants in the Los Angeles sample thus carry implications for what could happen among unauthorized Mexican immigrants and their children if new legalization programs and pathways to citizenship were legislatively adopted. The results in table 10.4 imply that the legalization and citizenship trajectories of those IIMMLA parents who started out as unauthorized migrants are importantly related to children's economic outcomes. For example, when parents who were initially unauthorized *changed* their legal status, and particularly when they also became naturalized citizens, a substantially reduced likelihood of educational failure is evident among their children. For example, about 35 percent *fewer* such children (those whose fathers initially resided as unauthorized immigrants but went on to LPR status and then eventually naturalized) failed to finish high school, compared with children whose fathers stayed unauthorized (14.5 percent vs. 22.5 percent for those whose fathers remained unauthorized; see table 10.4). In the case of finishing college, the children of unauthorized fathers who eventually naturalized graduated from college at almost twice the rate of children whose fathers remained unauthorized (19.1 percent for the former vs. 10.0 percent for the latter). The education gaps were even broader when we examine the children's schooling for mothers who changed status versus those who did not.

While the relative numbers of children going on to college in these instances is not inordinately high, the parents' migration status and citizenship trajectories clearly matter for children's well-being. In addition to these educational advantages, children of parents who were able to become LPRs, or who took advantage of the opportunity to legalize and then to

Table 10.4 Human Capital and Economic Attainment by Father's and Mother's Entry Status and Legal Status and Naturalization Trajectory, 1.5 and 2nd Generation Mexican-Origin Respondents

	Fathers			Mothers		
	All	Sons	Daughters	All	Sons	Daughters
Respondent Education						
% Less than High School Diploma						
Not Foreign-Born	11.8	17.0	5.0	15.3	12.5	17.2
Authorized / Naturalized	10.9	15.0	5.7	11.3	14.1	7.7
Authorized / Green Card	17.8	20.4	15.6	20.3	22.6	18.2
Unauthorized / Naturalized	14.5	11.8	17.1	11.6	12.5	10.8
Unauthorized / Green Card	20.2	18.2	21.4	18.3	17.2	19.0
Unauthorized / Unauthorized	22.5	20.0	25.0	35.9	39.1	31.3
Status Unknown	26.7	26.9	26.5	11.1	16.7	0.0
Never Lived in U.S.	37.0	40.7	33.3	49.4	50.0	48.7
% Bachelor's Degree or Higher						
Not Foreign-Born	19.4	22.6	15.0	8.2	7.5	8.6
Authorized / Naturalized	16.7	16.5	17.0	22.3	23.5	20.8
Authorized / Green Card	15.3	20.4	10.9	11.7	11.3	12.1
Unauthorized / Naturalized	19.1	22.4	15.8	20.3	21.9	18.9
Unauthorized / Green Card	14.9	11.4	17.1	11.3	10.3	11.9
Unauthorized / Unauthorized	10.0	5.0	15.0	0.0	0.0	0.0
Status Unknown	8.3	3.8	11.8	0.0	0.0	0.0
Never Lived in U.S.	8.4	10.2	6.7	8.6	11.9	5.1

Table 10.4 (Continued)

	Fathers			Mothers		
	All	Sons	Daughters	All	Sons	Daughters
Average Years of Education						
Not Foreign-Born	13.5	13.5	13.5	12.7	12.9	12.6
Authorized / Naturalized	13.3	13.2	13.5	13.7	13.5	13.9
Authorized / Green Card	13.0	13.0	13.0	12.9	12.9	12.9
Unauthorized / Naturalized	13.3	13.6	13.1	13.5	13.4	13.6
Unauthorized / Green Card	13.0	13.0	13.0	12.9	12.9	12.9
Unauthorized / Unauthorized	13.0	12.3	13.6	11.4	11.4	11.4
Status Unknown	12.4	12.3	12.4	13.0	12.7	13.7
Never Lived in U.S.	11.8	11.7	11.8	10.9	11.4	10.4
% Prefer to Speak English at Home						
Not Foreign-Born	71.0	73.6	67.5	80.6	87.5	75.9
Authorized / Naturalized	69.5	69.9	68.9	67.0	70.0	63.1
Authorized / Green Card	57.6	55.6	59.4	53.1	48.4	57.6
Unauthorized / Naturalized	61.8	65.8	57.9	55.1	64.1	47.3
Unauthorized / Green Card	36.0	38.6	34.3	37.3	34.5	39.3
Unauthorized / Unauthorized	45.0	40.0	50.0	28.2	65.2	25.0
Status Unknown	41.7	53.8	32.4	66.7	66.7	66.7
Never Lived in U.S.	32.8	32.2	33.3	28.4	33.3	23.1

Average Occupational Socioeconomic Prestige						
Not Foreign-Born	40.4	39.2	42.1	42.4	41.4	43.1
Authorized / Naturalized	42.1	40.8	43.8	42.5	40.8	44.8
Authorized / Green Card	42.9	41.5	44.1	40.3	39.5	41.1
Unauthorized / Naturalized	41.6	40.6	42.7	41.7	40.7	42.5
Unauthorized / Green Card	40.8	39.5	41.7	41.2	38.5	43.2
Unauthorized / Unauthorized	38.5	36.8	40.4	34.0	33.7	34.4
Status Unknown	39.6	37.4	41.4	42.6	39.3	49.2
Never Lived in U.S.	38.8	35.8	42.3	36.9	35.6	38.7
Average Personal Income						
Not Foreign-Born	23,194	26,962	18,200	25,847	24,363	26,871
Authorized / Naturalized	26,151	30,853	20,250	25,000	29,085	19,658
Authorized / Green Card	19,182	21,426	17,289	19,871	23,734	16,242
Unauthorized / Naturalized	23,638	28,480	18,796	21,960	25,445	18,946
Unauthorized / Green Card	20,061	24,000	17,586	18,123	22,948	14,792
Unauthorized / Unauthorized	16,988	18,850	15,125	14,218	17,587	9,375
Status Unknown	19,567	20,538	18,824	16,056	18,750	10,667
Never Lived in U.S.	17,395	19,169	15,650	19,685	25,833	13,064

naturalize, did better than with those with parents who remained unauthorized in terms of occupational prestige, income, and the tendency to speak English. The premium attached to having a mother who was initially unauthorized but then naturalized (vs. having a mother who remained unauthorized) was almost 15 percent for occupational prestige; about 50 percent for income; and over 95 percent for speaking English (table 10.4). Those whose fathers were initially unauthorized migrants but then went on to legalize as well as to become naturalized citizens reported average incomes of $23,638 in 2004. Those who had fathers, however, who were living in the country initially as unauthorized residents and then *stayed* unauthorized (i.e., were still unauthorized at the time of the IIMMLA interview in 2004) reported incomes that averaged only $16,988. In other words, on average, the former group made $6,650 more than the latter, or about 37 percent *higher* annual incomes—a considerable premium for having attained legal permanent resident status and having naturalized. The premium for those whose mothers became LPRs *and* subsequently naturalized versus those whose mothers remained unauthorized was even more notable: $7,742, or over 50 percent higher.

Did trajectory effects differ by gender? Here it is possible to examine either differential effects by the gender of the 1.5 or 2.0 children whose human capital or economic characteristics are being affected, or differential effects by the gender of the parents whose trajectories are being examined. In regard to the latter, the *mother's* legal status/naturalization trajectory appears to matter more than does the father's for children's outcomes (note the results in the "All" columns of table 10.4). Of course, mothers' and fathers' trajectories are often not independent of one another, but as a first approximation we treat them here without taking this into account. The difference in the tendency *not* to finish high school between children with mothers who stayed unauthorized and those whose mothers went from being unauthorized to naturalization is large (35.9 percent vs. 11.6 percent), compared with the same trajectory difference for fathers (22.5 percent vs. 14.5 percent). Thus, consistent with the idea that more traditional gender roles assign mothers the prime responsibility for socialization, the legal status/naturalization trajectories of mothers had a particularly strong impact on education and subsequent economic attainment among 1.5 and 2.0 generation children of immigrants.

But differences also emerged by gender of offspring. Father's trajectory makes more of a difference for sons than for daughters. This is what we would expect for an immigrant group coming from a country with more traditional gender roles and for a group with a very strong history of and orientation toward labor migration (i.e., migration occurring overwhelmingly for the purpose of working, often so that remittances can be sent back to family members in the sending country, with this labor migration role substantially concentrated among males). In short, fathers who initially start as unauthorized migrants and then go on to naturalize appear to impart advantages to their sons more than to their daughters, particularly in terms of earning a degree. While persistent unauthorized status among mothers seems to depress the income of offspring, the effect is particularly pronounced for daughters.

Discussion and Conclusions

The above results support the idea that parents' life trajectories influence children's life chances (Wagmiller et al. 2006). Here we find that the children of unauthorized Mexican immigrants who are able to change their legal status and naturalize show better socioeconomic outcomes than the children of immigrants who remained unauthorized. Such results perhaps provide a basis for gauging the long-term effects on immigrant group success of legislation that provides pathways to legalization and citizenship. Our findings suggest that the legalization possibilities made available by IRCA may have operated to enhance educational attainment, English usage, occupational prestige, and incomes on the part of the children of unauthorized immigrants. Pathways to legalization and citizenship thus seem likely to smooth the way for the children of unauthorized immigrants to become societal stakeholders in general. Conversely, the lack of such pathways risks raising the number of children growing up in poor and vulnerable households and adding to the size of any existing immigrant underclass. That the absence of legalization and citizenship pathways may also limit economic integration is further reinforced by research showing that Mexican immigrant parental legalization and citizenship raises the level of civic engagement in the case of their children, an outcome that generally also fosters economic success among immigrants (DeSipio, Bean and Rumbaut 2005).

Such ramifications of parents' legal status apply to both fathers and mothers, although relationships observed here are also gendered in interesting ways. Mothers' status has the greater effect on children's education. The level of legalization and naturalization among both fathers and mothers is high, with this similarity holding over the course of several decades after the parents' arrival in the United States. However, mothers' trajectories, particularly when they lead from unauthorized status to naturalization, appear to influence daughters more than sons. This finding is consistent with the implications for socialization specialization attached to relatively traditional gender roles.

It is also worth noting that the above results do not mean that the parents' legalization and citizenship per se *cause* children's higher economic attainment, although they may, in part, particularly by improving access to economic opportunities available only to legal immigrants or citizens. It is possible that such results derive, in part, from processes of selectivity. That is, it may be the case that the smartest and most industrious of the parents are also more likely to legalize and obtain citizenship, and that the influence of such tendencies helps to account for the gains in education and income among their children rather than the effects of legal and citizenship status change after they occur. However, even if this were the case, it would not diminish the significance of the structural opportunities provided by legalization and citizenship pathways. The reason is that the presence of such opportunities, including even the *prospect* that they may emerge, is a prerequisite for positive selection to occur (Heckman 1992, 1996, 1997). Moreover, this is precisely the intention of public policies encouraging and allowing for legalization and naturalization (Heckman, Smith and Clements 1997). That is, positive selection in this instance is *not* something merely to be accounted for statistically, but rather is an important substantive process. Also, without legalization and citizenship opportunities, the migration of the motivated and industrious might be substantially reduced. This itself could contribute to the further development of an impoverished, vulnerable, and perhaps alienated underclass of unauthorized migrants in the United States. That the chance to become full members of society matters is indicated by the fact so many immigrants, particularly Mexican immigrants, legalized and became citizens as a result of IRCA when they were presented with the opportunity to do so.

If legalization and citizenship programs had been unavailable at the time, it is unlikely the parents in the IIMMLA sample would have fared as well in America as they did. They would have had to live and work in the shadows to a much greater degree, and in all probability would have lacked the resources to provide as well for their children, including the resources to help pay for college. Without the possibility to legalize and become citizens (i.e., in the form of signals of both a welcoming social reception and opportunities to legalize), they would in all likelihood have been less motivated to succeed than they were (Van Hook, Brown and Bean 2006). Overall, migration and citizenship opportunities appear to matter considerably. By providing environments that encourage educational attainment and economic achievement among the children of immigrants, legalization and citizenship pathways are likely not only to facilitate the economic integration of the immigrant generation, but also the 1.5 and second generation as well.

Acknowledgments

Appreciation is expressed to the Russell Sage Foundation, which underwrote the collection of data for the Immigration and Intergenerational Mobility in Metropolitan Los Angeles (IIMMLA) study, and to the Hewlett Foundation, which supported the collection of data on Mexican immigrants in Los Angeles. Infrastructure assistance from the Center for Research on Immigration, Population and Public Policy at the University of California, Irvine, is also gratefully acknowledged.

Note

1. All of the results presented in this chapter are statistically significant. Because the legalization and citizenship trajectories involve combinations of transition points (i.e., a respondent cannot be classified as having a parent with certain trajectories unless that parent happened to have passed certain previous transition points), it is appropriate to consider the entire set of trajectories together in conducting tests of statistical significance. Thus, in the cases of each of the human capital or labor market outcome variables examined (and for both parents and both genders of offspring), the set of trajectories examined was statistically significant when either conventional analyses of variance for trajectory-set differences were conducted or latent class analyses conducted for the significance of differences among "bundles" of trajectories from the set. That is, the differences in human capital or labor market outcomes across the statuses and trajectories (or their latent class combinations) would not be expected to have occurred by chance except at levels of probability less than one in 20.

Part IV

Educating Immigrants

To help young immigrants succeed in school is a crucial step toward helping them to succeed in life. The chapters of Part IV, therefore, focus on a particularly salient feature for the young immigrants' future—education.

The chapter by Carola Suárez-Orozco and Francisco X. Gaytán examines the schooling experiences of recently arrived young immigrants from a variety of countries. The authors confirm the paramount importance of English language fluency in predicting academic performance; it is the single most powerful student-level predictor of academic outcomes. They note critically that there is a dearth of policies for optimally educating immigrants and easing their transition to the labor market or college, and they make several policy conclusions based on their research, such as emphasizing the need for after-school programs and mentoring.

In his chapter, Joel Perlmann examines whether educational interventions should focus on raising high school or college graduation rates for Mexican American males (i.e., American-born sons of Mexican immigrants). The author concludes that greater benefits can be obtained by boosting the Mexican American high school completion rates than by boosting college completion rates. He further argues that such a policy focus on high school completion might also counteract the development of a large Mexican American underclass.

Robert Smith's chapter, discussing the efforts by CUNY and other educational institutions in New York City to improve the education of young Mexican immigrants, provides a transition to the concluding section of this volume, which focuses on aid organizations that provide concrete assistance to young refugees and immigrants.

11

Schooling Pathways of Newcomer Immigrant Youth*

Carola Suárez-Orozco and Francisco X. Gaytán

Outside the family, schools are the most important context of social development shaping the lives of newcomer immigrant youth. It is the first sustained, meaningful, and enduring site of participation in an institution of the new society (Suárez-Orozco and Suárez-Orozco 2001). It is in schools that immigrant youth begin to acquire the academic, linguistic, and cultural knowledge necessary for their success in the United States. Immigrant students, new to the American system, rely heavily on school personnel—teachers, counselors, coaches, and others—to guide them in the steps necessary to successfully complete their schooling and, with luck, to go on to college. It is through their interactions with peers, teachers, and school staff that newly arrived immigrant youth experiment with new identities and learn to calibrate their ambitions (Stanton-Salazar 2004). These relationships serve to shape their characters, open new opportunities, as well as set constraints to future pathways. It is in their engagement with schooling, broadly defined, that immigrant youth most profoundly transform themselves.

In this chapter we present data from the Longitudinal Immigrant Student Adaptation Study (LISA), a five-year study that combined interdisciplinary and comparative approaches to document patterns of adaptation of recently arrived immigrant youth from China, Central America, the Dominican Republic, Haiti, and Mexico. We map 309 immigrant students' patterns of academic engagement and performance over time, provide explanatory models, and end by presenting recommendations for policy and practice to better serve immigrant youth.

The Long View on Academic Performance

As students move from elementary school into middle school and as they transition into high school, research has demonstrated a pattern of decline in motivation and performance in school for all youth—immigrant and nonimmigrant, white, black, Latino, or Asian (Eccles, Wigfield, and Schiefele 1998). This pattern has been found to be strikingly precipitous among minority (Fredricks, Blumenfeld, and Paris 2004) as well as immigrant students (Steinberg, Brown, and Dornbusch 1996).

Our data support results reported in earlier studies. On an average, the longer the newcomer immigrant students in our study were in school, the worse they achieved academically as measured by grade averages, steadily dropping from a GPA of 2.89 Year 1 (1998) to 2.42 Year 5 (2002).

Country of Origin Patterns. With the exception of the Chinese students, the grades of all the groups in our sample declined over time. While the grades of Dominicans and Central Americans dropped three-fourths of a grade, the grades of the Haitians participants declined slightly over half a grade, and those of Mexicans dropped slightly over one-third of a grade between the first and fifth year of the study. To understand what contributes to academic success across different immigrant groups, we utilized latent class growth modeling (Nagin and Tremblay 1999) to determine distinct trajectories of performance across groups.[1] Five performance pathways emerged—consistently high performers (*High Performers*), consistently low performers (*Low Performers*), students whose GPA slowly drifts downward across time (*Slow Decliners*), another that declined in precipitously in academic performance (*Precipitous Decliners*), and a group that improved over time (*Improvers*).

Disconcertingly, two-thirds of all the participants in this study demonstrated a decline in their academic performance over the course of five years. Nearly a quarter of the students were Slow Decliners (GPA decline of half a grade over five years). Another quarter of the participants were Precipitous Decliners (GPA decline of more than a whole grade). Further, 14.4 percent were Low Performers throughout the course of the study; they began with a lower average GPA than any of their peers and dropped an additional half grade by the study's end.

By contrast, two groups of students demonstrated performance that defied the pattern of decline. Nearly a quarter of the students in the sample were High Performers maintaining an average GPA of 3.5 across the five years of the study. The last group of Improvers made considerable strides in augmenting their performance by a little more than two-thirds of a grade over time to a respectable B average.

Factors Influencing Academic Achievement

What factors influenced these academic trajectories? To explain the academic performance of immigrant newcomer students, we considered a number of factors that have been associated with academic performance in the research literature (see table 11.1, *Correlation Matrix*).

Family Structure. It is well established that children who grow up in two-parent families tend to be at an advantage academically (Astone and Mclanahan 1991, Boyce Rodgers and Rose 2001). Immigrant children often have a variety of familial and household configurations due to separations, living with extended kin, and blended families. We found that having two or more adults in the home had a weak but significant positive relationship to grades.

Parental Education. There is a close relationship between parental education and higher performance on achievement tests and grades, better school readiness, lower dropout rates, lower levels of school behavior problems, and higher school engagement (Jencks 1972, White 1982, Sirin and Rogers-Sirin 2005). Our study shows that both father's education and mother's education had a weak but statistically significant positive association with grades as well as achievement test scores (see table 11.1). Further, our results show a significant association between father's education and school quality measures (discussed later), including percentage of students performing at proficiency in state tests, and a negative correlation with school segregation and poverty.

Parental Employment. Parents who are active in the labor force and who thus generate income are better able to buffer their children from the risks associated with poverty. We found a positive association between father working and GPA as well as test scores. There was no association, however, between the mothers working outside the home and the grades or test scores the children received.

Gender. Girls and young women are now outperforming boys and young men on a variety of academic measures (Brandon 1991, Portes and Rumbaut 2002, Suárez-Orozco and Qin-Hilliard 2004). A number of factors have been identified in contributing to this phenomenon. Girls are more likely than boys to comply with the often tedious behaviors that are expected in classroom settings, while boys are also more likely to engage in disruptive behaviors (Kenny-Benson et al. 2006). In many

Table 11.1 Correlation matrix

Variables	M	SD	1a	1b	2	3	4	5	6	7	8	9	10a	10b	10c	10d	11	12	13a	13b	14	15	16
1a. Outcome: GPA Year 5	2.43	.89	1																				
1b. Outcome: Woodcock-Johnson Combined Reading/Math	90.60	16.47	.48***	1																				
2. Behavioral Engagement	20.89	4.22	.42***	.18**	1																			
3. English Proficiency	75.06	19.12	.41***	.86***	.22***	1																		
4. Cognitive Engagement	11.75	1.92	ns	ns	.32***	ns	1																	
5. Relational Engagement	36.50	4.90	ns	ns	.35***	.41***	.42***	1																
6. Wellbeing	18.63	9.81	−.16**	ns	−.21***	ns	ns	−.20***	1															
7. Academic Self-Efficacy	19.22	2.33	ns	ns	.18**	ns	.30***	.23***	ns	1														
8. Attitudes Towards School	5.99	1.39	.27***	.17**	.35***	.23***	.36***	.41***	−.20**	.19**	1													
9. Perceptions of School Violence	20.23	5.58	−.24***	−.27***	−.32***	−.23***	−.14**	−.20**	ns	ns	−.29***	1												

	Mean	SD															
10a. School: ELA Level 3 or 4	31.57	25.29	.28***	.52***	.20**	.48***	−.17**	ns	ns	−.14*	ns	−.35***	1				
10b. School: Non-White	78.84	23.30	−.29***	−.44***	−.19**	−.41***	.20***	ns	ns	ns	.30***	−.77***	1				
10c. School: % Low Income	49.64	23.89	ns	−.27***	−.18**	−.28***	.13*	ns	ns	ns	.28***	−.62***	.77***	1			
10d. School: Attendance Rate	92.53	5.17	.22***	.26***	ns	.28***	ns	ns	ns	ns	.19**	−.20**	ns	1			
11. Two Parent Household	68%	n/a	.23***	.17*	.18**	.12**	ns	ns −.13**	ns	.14* −.17*	ns	ns	.14*	1			
12. Female	58%	n/a	.21***	ns	.21***	ns	.11*	.22***	ns	ns	ns	ns	ns	ns	1		
13a. Maternal Education	33%	n/a	.13*	.25***	.13*	.26***	ns	ns	ns	ns	.19**	−.12*	ns	ns	ns	1	
13b. Paternal Education	24%	n/a	.12*	.20***	ns	.25***	ns	ns	ns	−.22***	.265***	−.26***	−.21***	ns	ns	.29*** ns	.35*** 1
14. Paternal Employment	65%	n/a	.21***	.13*	.17**	ns	ns −.15*	ns	ns	ns	ns	−.13*	ns	ns	.70*** ns	.28*** 1
15. Age at Arrival	9.83	2.00	ns	−.22***	ns	.22***	ns	ns	ns	ns	ns	ns	−.12*	−.26***	ns	ns	ns 1
16. Years in the U.S. (at year 5)	6.91	1.38	ns	.17**	ns	.22***	ns	ns	.12*	ns	ns	ns	.14*	ns	ns	ns	ns −.59*** 1

Note. # We also examined the correlations of two additional predictors to our outcomes of interest—age and maternal employment, which were both nonsignificant.
For these analyses we used only those students for whom we had complete data: N = 284. Maternal Education & Paternal Education (Completed High School = 1, Did Not Complete High School = 0), Paternal Employment (Father Works = 1, Father Does Not Work = 0).
* < p .05; ** < p .01; *** < p .001

settings, boys are subjected to higher levels of physical intimidation and aggression than girls. Boys of color face lower expectations, more stigmatization, and are subject to more blatant discrimination than girls and are thus at greater risk for academic disengagement (Crul and Doomernik 2003, López 2003). In addition, as immigrant girls often have many more responsibilities at home than their brothers, they may be predisposed to finding school a more engaging social space (Fuligini and Pederson 2002). Immigrant boys tend to be allowed more freedom, which may render them more susceptible to lure of the street (Waters 1996, Olsen 1997, Valenzuela 1999, Sarroub 2001). In addition, boys tend to have fewer meaningful relationships with their teachers and perceive their school environments to be less supportive than their sisters do (Suárez-Orozco and Qin-Hilliard 2004).

As expected, we found that girls in our sample had higher grades and substantially outperformed boys in every immigrant-origin group for the duration of the study. No gender differences were found in standardized test scores, however. This discrepancy may be related to the relative importance of the behavioral component in predicting GPA (discussed later).

School Contexts. Many immigrant children, especially those who live in poor urban neighborhoods, face daunting odds in their schools and communities (Waters 1999). Neighborhoods characterized by greater levels of unemployment (Wilson 1997), violence, barriers to access (Massey, and Denton 1993), and intense segregation by race and poverty (Orfield and Yun 1999, Orfield and Lee 2006) tend to have schools that are overcrowded and understaffed, face high teacher and staff turnover, are poorly resourced, maintain low academic expectations, and are plagued by the ever-present threat of violence and hostile peer cultures (Mehan et al. 1996, García-Coll, and Magnuson 1997). School contexts matter in the engagement and performance of students (McLaughlin, and Talbert 1993, Samdal et al. 1998, Bryk and Schneider 2003).

Since students of color and those attending urban schools are most likely to encounter violence, such concerns affect a disproportionate number of immigrant students. We found that students' perceptions of violence in their schools and neighborhoods were negatively correlated to academic performance (GPA and achievement scores), relational engagement, cognitive engagement, behavioral engagement, and the level of English Proficiency.

We also examined the relationship between the students' perceptions of school violence and school district measures of school problems: segregation rate, poverty rate, attendance rates, and the percentage of students in the school passing the state-mandated high-stakes English

Language Arts (ELA) test. Each of these characteristics has been associated with student performance (Slotnik and Gratz 1999, Hulpia and Valcke 2004). Our data show a positive correlation between the students' perceptions of school problems and the school's proportion of nonwhite students and low-income students and its average daily attendance rate, as well as a negative relationship with the school's ELA test passing rate.

English Language Proficiency (ELP). Language skills affect students' abilities to detect social nuances in the school setting and are also highly predictive of academic success (Muñoz-Sandoval et al. 1998). While verbal proficiency can be developed within a couple of years, the level of language skills necessary to be competitive with native-born peers in the classroom takes on an average five to seven years to acquire under optimal conditions (Cummins 1991, Collier 1992, Klesmer 1994).

Consistent with the evidence of this line of research, in the LISA study, higher ELP was significantly related with higher grades and even more strongly related to achievement test outcomes. Length of residency in the United States, maternal education, and the student's perceptions of school violence as well as the school district indicators of quality (high-stakes state test of ELA proficiency rate, segregation rate, poverty rate, and daily attendance rate) were all significantly correlated to ELP (Carhill, Páez and Suárez-Orozco 2008).

Next, we discuss attitudinal variables that are likely to predict a student's educational trajectory.

Well-being influences a student's ability to focus on their studies. A stressful school climate, characterized by perceptions of academic pressure, danger, discrimination, and the absence of supportive relationships, can undermine students' well-being, taxing their abilities to cope (Karatzias et al. 2002). Conversely, a more supportive educational context can have a protective effect on students' well-being (Samdal et al. 1998).

Academic Self-Efficacy is the belief that one is competent and in control of one's learning. It has been demonstrated to predict the extent to which a child engages in learning the new language, forges new relationships, and connects with the academic tasks at hand (Schunk 1991). While greater academic self-efficacy was not directly linked to either grades or test scores, it *did* have a significant relationship to both cognitive and relational engagement.

Attitudes Toward School of immigrant students have been shown to be more positive than those of their native-born peers (Kao and Tienda 1995, Suárez-Orozco and Suárez-Orozco 1995, Fuligni 1997). We found attitudes toward school to be related to grades, and to have even stronger associations with cognitive and relational engagement.

Academic Engagement

We further considered the key concept of academic engagement. This concept describes the extent to which students are connecting to what they are learning, how they are learning it, and with whom they are learning. These factors appear to play a central role in how well they do in school (Greenwood, Horton, and Utley 2002, Fredricks, Blumenfeld, and Paris 2004, National Research Council 2004). We distinguished three dimensions of academic engagement—cognitive, relational, and behavioral.

Cognitive Engagement. Cognitive engagement is the degree to which students are interested in and curious about what they are learning. Curiosity can serve to engage students in their learning experiences. Our cognitive engagement scale examined the degree to which the students are engrossed and intellectually engaged in what they are learning (e.g., "I enjoy learning new things").

Relational Engagement. Relational engagement is the extent to which students feel connected to their teachers, peers, and others in their schools. Successful adaptations among immigrant students appear to be linked to the quality of relationships that they forge in their school settings (Cauce, Felmer, and Primavera 1982, Dubow 1991, Wentzel 1999). Social support in the school has been implicated in the academic adaptation of all students, and immigrant students appear to be no exception (Zhou and Bankston 1998, Portes and Rumbaut 2001).

Behavioral Engagement. We view behavioral engagement as a component of academic engagement that specifically reflects students' participation and efforts in academic tasks, such as expending their best effort in completing class work and homework, turning in assignments on time, paying attention to class work, classroom behaviors, and attendance.

Longitudinally, we found a distinct pattern of accelerating behavioral disengagement for the sample as a whole. Interestingly, there was little difference in the amount of effort boys and girls expended in their studies initially. Over time, however, girls maintained their levels of behavioral engagement while boys were more likely to disengage. As we expected, the consistent high performers were significantly more behaviorally engaged in school than either the students who took low or declining performance pathways.

Our data reveal interesting relationships among these dimensions, as well as between them and various predictor and outcome variables (see table 11.1). In this chapter, we have to forego a detailed discussion and refer the interested reader to the chapter "Networks of Relationships" in Súarez-Orozco, Súarez-Orozco, and Todorova (2008).

Predicting Academic Achievement

How well do the predictors in our conceptual model explain academic performance?

Predicting Grades. We began by examining the impact on GPA of student-centered characteristics, behavioral engagement, ELP, having two parental figures in the home, maternal education, and a father who works—all factors implicated in the research literature and by our modeling of academic engagement to predict academic achievement (see table 11.2). We found that grades were positively correlated to all of these variables.

The most robust predictor of GPA was ELP, followed by behavioral engagement. Students with higher levels of English skills were more likely to earn better grades. Further, the more the students engaged in the behaviors necessary to do well in school, the more likely they were to attain higher grades. Other student background characteristics (such as mother's education and whether or not the student came from a two-adult household) were not significantly related to better grades in this model. This student model of academic achievement, focusing on the student characteristics and perspective, provides substantial insight into predicting student grades.

Predicting Achievement Test Scores. To explore a more "objective" outcome of performance, we selected the same five variables as predictors that we used to predict GPA—behavioral engagement, ELP, having a working father, maternal education, and having a two adult figure household (see table 11.3). We substituted grades with a combined score derived from four subtests of the Woodcock-Johnson Test of Achievement-R (WJTA-R), which we had individually administered to our participants.

Table 11.2 Predicting GPA—Student Centered Perspective

Independent Variable	B	S.E	t	p
Intercept	−0.273	0.271	−1.01	ns
English Proficiency	0.0151	0.00256	5.9	***
Behavioral Engagement	0.0655	0.0114	5.72	***
Two-Adult Household	0.161	0.139	1.16	ns
Working Father	0.132	0.137	0.97	ns
Mother's Education	0.00907	0.103	0.09	ns
$R^2 = 29.67$, $F(5, 261) = 22.03^{***}$				

Note: Multiple Regression Analysis Describing the Relationship of Student's GPA at Year 5 to Student's English Language Proficiency (ELP), Behavioral Engagement, Household Composition (two adults), Father's Employment Status, and Maternal Education.

Table 11.3 Predicting Standardized Achievement Test—Student-Centered Perspective

Independent Variable	B	S.E	T	p
Intercept	34.19	3.12	10.96	***
English Proficiency	0.732	0.0291	25.2	***
Maternal Education	1.0272	1.174	0.87	ns
Two-Adult Household	2.67	1.571	1.7	ns
Behavioral Engagement	−0.034	0.13	−0.26	ns
Father work status	0.157	1.544	0.1	ns
$R^2 = 74.14$, $F(5, 259) = 148.248$***				

Note: Multiple Regression Analysis Describing the Relationship of Student's Standardized Achievement Test at Year 5 to Student's English Language Proficiency, Behavioral Engagement, (ELP), Household Composition (two parental figures), Father's Employment Status, and Maternal Education.

Of these factors, ELP again explained the most variance by far—the better a student's academic English proficiency, the better they were able to do on a standardized achievement test. This is not surprising given that the achievement test is highly dependent upon how well a student is able to manage and manipulate academic English; two of the four subtests that make up the Broad Reading and Math achievement test score require sophisticated understanding of English (such as the Math Word Problems subtest and the Reading Comprehension subtest). All of the other variables we examined in this model—behavioral engagement, maternal education, having a working father, and having a two-parent family—did not significantly contribute to explaining the achievement test outcome.[2]

Because student level factors explained so little of the achievement test outcome (once the ELP score is disregarded), we turned to examining school level factors (see table 11.4).[3] The best predictor of the achievement test outcome was the percent of students attending the school who performed at proficiency or above on the state's mandated ELA exam. The lower the average daily school attendance rate, the less well our participant was likely to do on the achievement test. Further, the poorer the students in the school and the more racially segregated the school the newcomer student attended, the less well our participants were likely to do on the achievement test.

Implications for Intervention

These findings have implications for practice and intervention. We used a variety of student-focused as well as school-focused variables to predict

Table 11.4 Predicting Standardized Achievement Test—School Perspective

Independent Variable	B	S.E	T	p
Intercept	42.357	17.134	2.47	*
ELA Proficiency Rate	0.299	0.0541	5.54	***
School Daily Avg. Attendance Rate	0.478	0.17	2.81	**
School Poverty	0.127	0.0568	2.24	*
School Racial Segregation	−0.151	0.0701	−2.15	*
$R^2 = 32.19$, $F(4, 256) = 30.38^{***}$				

Note: Multiple Regression Analysis Describing the Relationship of Student's Standardized Achievement Test Score at Year 5 to ELA Proficiency Rate, School Attendance Rate, and School Racial Segregation, School Poverty.

two academic outcomes—GPA and achievement tests. Our analyses show that while the 'usual suspects' of background structural characteristics such as parental education, having a two-parent family, and family income contribute to academic performance, several other factors provide far greater explanatory power. It would be fair to conclude that *ELP is the single best student-level predictor of academic outcomes* as measured by an achievement test as well as GPA. English language proficiency had triple the predictive value of all the other student variables combined for the achievement test score and played a significant role in grades. The significance of school contexts in academic performance cannot be underestimated. Further, at the individual level, behavioral engagement—which was heavily influenced by relational and cognitive engagement—was a very robust contributor to better grades. Lastly, quality of school context variables highly influenced both forms of academic performance—grades and standardized achievement test measures.

Years of basic research in schools serving large numbers of immigrant youth have made it painfully obvious that we have no coherent policies for optimally educating immigrants—especially immigrants arriving during the middle and high school years (Ruiz-de-Velasco et al. 2001)—nor do we have any significant frameworks to ease their transition to college or to the labor market. What we have now in place is a policy of "non-policy"—a sink or swim approach that seems to rest on the faith that after immigrants cross the border, the logic of the market and the magic of American culture will work to somehow turn them over time into proud, loyal, and productive citizens.

Today's globally linked economies and societies are unforgiving of those without the higher-order cognitive and meta-cognitive skills, cultural sophistication, and ability to manage complexity that are imparted in our better secondary and tertiary educational settings (Suárez-Orozco and Qin 2004). The Gates Foundation (2006) argues that the new economy requires schools to provide "new 3 Rs"—*rigor* in challenging classes; *relevance* to engaging topics that "relate clearly to their lives in today's rapidly changing world"; and *relationships* with adults "who know them, look out for them, and push them to achieve," though far too few immigrant students attend schools that provide these experiences. Further, middle-class Americans understand nonservice sector jobs require a college education, although too many immigrant youngsters receive a mediocre high school education that leave them out of the college pathways.

Most of the students in our study attended high-poverty, highly segregated schools where more than three-quarters of their peers were of color. In recent years, despite legislation in the 1960s designed to reduce segregation, our schools have become increasingly resegregated. This new pattern of segregation tends to be not just about color but also is accompanied poverty, and linguistic isolation—so-called triple segregation (Orfield and Lee 2006). This is deeply troubling as these types of segregation have proven to be inexorably linked to negative outcomes—including climates of low expectations and academic performance, reduced resources, lower achievement, greater school violence, and higher dropout rates.

In such settings, students' opportunities and experiences are limited in a variety of ways. Resources are scarce, and often the rundown buildings reflect this. Classrooms are typically overcrowded and the curriculum is outdated and irrelevant. Classroom routines can be unengaging—often consisting of below-grade-level worksheets or outdated videos. Instruction is often tracked, especially for English Language Learners who are put into "dummied down" classes in which they fall further and further behind their English-speaking peers. Few students go on to four-year colleges; hence college counseling is minimally available. In such settings, teachers tend to be inexperienced, teaching outside their content area or un-credentialed. Morale is low and teacher and principal turnover is high.

Nowhere is our policy and thinking about immigration more disconnected from the realities in the field and self-defeating than in the area of language. In the era of global societies and economies, multilingualism is a decisive asset, not a divisive threat. Yet we are doing nothing to support heritage languages and we are doing little to intelligently support students in the systematic development of academic English. Alarmingly, we found in nearly every school setting we investigated (over 100 during the course of the study) nearly universal sense of alienation within schools between the language programs for immigrant students (including various versions of bilingual education or English as a Second Language teaching) and the mainstream programs.

In many schools the separation between the immigrant English language learners and their mainstream peers was total and hermetic. As a result of this general trend we found few meaningful exchanges and friendships between newly arrived immigrant youngsters and native-born students. The *status quo* is generating innumerable missed opportunities: immigrant youngsters have little exposure to the linguistic modeling their American-born peers could provide, and American students, in desperate need to learn about the world beyond our borders, are missing out also by generally not connecting with their immigrant peers. Thus, there is little opportunity to practice and develop their new language skills.

Our data reveal that learning the academic English needed to do well in school, and ultimately in the workforce, takes much more time than our impatient policy makers would like. Immigrant youth may be able to pick up enough English to discuss with ease the latest ball game, video-game fad, or fashion frenzy. Yet when we systematically tested their verbal abilities in more complex domains, the data reveal a sobering pattern of slow gains in academic English. Hence, current proposals that would induce newly arrived immigrant students to the high-stakes testing regime of No Child Left Behind (NCLB) after just one year in the United States are unrealistic. Our data suggest that the majority of immigrant children cannot possibly be expected to master the complex intricacies of academic English in one year of study, particularly in the highly dysfunctional schools where huge numbers of newly arrived immigrant students concentrate. We witnessed many immigrant youths check themselves out of the schooling game as a result of their multiple failures in mandatory high-stakes testing—a tragedy for the child and loss for society.

Sadly, we found that the vast majority of recently arrived immigrant youth were quite isolated as they left school. Very few reported any kind of structured afterschool activities or connection to community organizations or community adults. In exploring immigrant students' time outside of school we disconcertingly found that only a small portion—7.4 percent—of the participants in this study reported spending regular time in community centers. Only 9 percent of the students in this study reported involvement in after-school classes or academic enhancing programs such as Upward Bound.

After-school (Noam, Miller, and Barry 2002) and tutoring programs can be of critical importance to immigrant students in providing them with homework help. Immigrant families by-and-large could not provide the kinds of academic supports that enhanced their performance. Though immigrant families held high educational aspirations for their children, they were often unable to help with homework or provide informed advice on accessing college (see Smith in this volume).

In response to the question about who provided them with support and guidance, we found that none of the students in our sample named a mentor provided by a formal mentoring organization. Only 14 percent of the total sample named a nonfamilial adult—3 percent named a community leader and 9 percent named someone in a school. Notably, high achievers were twice as likely as the other students to have developed a supportive relationship with a nonfamily member.

We asked students to tell us who they were most likely to turn to when they needed academic supports—help with homework, expecting good grades, talking to them about the future, and providing information about getting to college. Thirty-eight percent relied on their family for *help with homework*, 13 percent found help from people in school or after-school activities, and 16 percent reported having no source of help with homework. Forty percent reported that they received information about how to get to college from a family member; only 28 received this information from a logical and likely to be more knowledgeable source—counselors and teachers in school or at after-school sites. Alarmingly, 20 percent of the participants reported having no source of information about college access whatsoever.

Thus, a domain that is ripe for creative policy work that could help immigrant families enormously is the area of community supports and mentorship. The evidence suggests that such programs, when carefully planned and well staffed, can help all children and may play an especially important role in the lives of disoriented new arrivals in need of direction. Mentors can act as cultural guides to help them find their way during the turmoil years of adolescence in a new country. Caring adults can serve to illuminate the labyrinth to college pathway access that native born middle class parents find daunting and newcomer immigrants find incomprehensible. Since immigrant boys, our data reveal, encounter more serious difficulties in their journey to the new land, mentorship opportunities could be designed with their specific needs in mind. Behind nearly every successful immigrant youth journey we found a mentor—from the church, the athletic team, or from the local community center, take a youngster under his or her wings. But in nearly every case we encountered, mentorship relationships started by chance encounters.

We need a much more coordinated effort to link immigrant youngster who wants with a caring and knowledgeable mentor. Refugee organizations have decades of experience providing supports to newcomer youth who arrive as refugees. Many of the needs of newcomer immigrant youth are quite similar to those of refugees—counseling to aid with disorientation; language training; and access to jobs, housing, and health care among others. Scaffolding on the lessons learned and infrastructure of such organizations as Bridging Refugee Youth and Children's Services, Lutheran Immigration and Refugee Service, and Jewish Family Services (among others) is a sensible and an efficient way to begin to meet these kinds of orientation needs of newcomer, non-refugee immigrant youth.

Our society must develop more coherent education policies to ease new immigrant students' path to college and their eventual transition to the knowledge intensive economies of the global era. Policies matter—countries that better coordinate their migration objectives with proactive integration policies—including language and education policies—tend to have better outcomes in terms of the long term adaptation of immigrant youngsters as measured by better schooling performance (OECD 2005). Schools in urban America today, especially in cities that are being transformed by large-scale immigration, ought to be the very frontlines of an ambitious new incorporation agenda. With immigrant-origin youth the fastest growing sector of the U.S. child population, there is much at stake for the future of our economy and society.

Notes

*Parts of this chapter were first developed for: Suárez-Orozco, C., Suárez-Orozco, M., and Todorova, I. (2008). *Learning a New Land: Immigrant Students in American Society.* Cambridge, MA: Harvard University Press. Details of the sample characteristics, methodological strategy, and findings are available in this book.

1. This strategy identifies clusters of individuals on the basis of developmental trajectories (Nagin and Tremblay 1999) to establish the number of groups that best fit the data utilizing both patterns of individual change and probability of group membership The key outputs of the model estimation consider the shape of the group's trajectory, the estimated proportion of the population that belongs to each trajectory, and the probability that an individual would belong to each group. This semi-parametric approach does not make the assumption that parameters are normally distributed in a population (Nagin and Tremblay 1999). Several models with a variety of groups are tested; a determination is then made about the number of groups that best describe the data. It is recommended that the model with the largest Bayesian Information Criterion (BIC) be selected. We tested several models ranging from three to six categories. The model with five categories had the highest BIC and was thus selected to describe the data.
2. As it was clear that how developed a students' English proficiency was played such a clear role, we ran the same student-focused model *excluding ELP*. While

we had accounted for *74 percent* of the variance of the achievement test outcome when we included the ELP score, we accounted *only for 10.7 percent* of the variance when we excluded that predictor from our model.
3. Note that for comparison's sake, we ran this same School Perspective model, substituting GPA as the outcome measure. The model worked much less well for GPA, accounting for only 15 percent of the variance ($R^2 = 15.07$, $F(5, 258) = 11.45^{***}$).

12

The Importance of Raising Mexican American High School Graduation Rates

Joel Perlmann

Introduction

I discuss in this chapter the relative importance of two different social policy interventions to improve educational attainment. Specifically, interventions could focus on raising high school or college graduation rates. The issue of which intervention to call for was brought home to me in a seminar about my recent book, *Italians then, Mexicans now* (Perlmann 2005). To some extent, my arguments are particular to the Mexican second generation today, because their high school dropout rates are distinctly high. But to the extent that other groups' rates approximate those of the Mexican second generation, the point will be true for these other groups as well. In any case, whatever is true for the Mexicans is critical to know because they are by far the largest single immigrant group, and Mexicans are an even larger proportion of low-skill immigrants.

The points I make are simple: we don't want to ignore high school dropouts, and obviously we want to stimulate college graduation. Yet there is nothing like observing how a simple truth can be ignored to make one wish to repeat it. I first restate a formulation from my book that drew a troubled comment from a colleague. Then I present his concern and offer some general comments about it. And finally, I offer a new, more specific, model of what can be expected under different assumptions about educational change among second generation Mexican Americans.

Education and Earnings: Native-born Mexican, African American and White American Men, 2000

I studied the education and earnings of the Mexican second generation (relying on the 2000 Census microlevel data).[1] I estimated that high school dropout rates for the group were alarmingly high—around 25–30 percent for young adult men. These high rates may not surprise you because we often hear reports about the high proportion of young adult Hispanics who are not high school graduates. But typically those figures fail to distinguish between the native born and the immigrant, for example, in the reports of the Department of Education. And so the figures about high Hispanic dropout rates with which we are familiar are largely useless as a measure of immigrant absorption or the extent of ethnic inequality among the American born. Rather, such figures simply tell us that Hispanic immigrants are a large proportion of all young adult Hispanics, and that many Hispanic immigrants have failed to complete high school, typically having dropped out while still living in their country of origin. That is important to know, but it is not the measure we want in order to understand the success of American schools in reaching Hispanics. By contrast, the figure I am reporting is so disturbing because it is based on a much more refined measure, limited to the American-born children of Mexican immigrants.

The Mexican second generation rate for high school dropout is twice the rate found among native black men of the same age, and the comparison with non-Mexican native whites, of course, is even more extreme. At the same time, rates of college completion for second generation Mexican men *who have finished high school* are also far lower than the rates of college completion for native whites who have finished high school. In short, for the Mexican second generation, there are two distinct forms of educational vulnerability involved in low college completion: low high school completion rates and low persistence in education through four more years by high school completers. By contrast, only the second form of vulnerability has an important impact on white-black educational attainment differences.

How much improvement in Mexican second generation earnings might we reasonably expect if either of these levels of vulnerability were eliminated? Even if *all* educational differences between native white and the Mexican second generation men were eliminated, only *half* of the ethnic earnings gap would disappear—the ethnic earnings ratio rising from 75 percent to 87 percent. So ethnic inequality is certainly not just a story about schooling. Nevertheless, with schooling accounting for half the ethnic inequality, we do want to know what impact is created by the Mexican failure to complete particular *levels* of schooling at the rates that native

Table 12.1 Modeling the improvements in ethnic earnings that would result from improvements in ethnic educational attainment, men 25–34 in two groups

Advantages in earnings gained expressed as a proportion of the entire earning gap related to education *when each group is compared with native whites*

	Mexican Second Gen.	native black
Scenario 1. Each group reaches native white educational attainments	**1.00**	1.00
Scenario 2. The percentage graduating from high school remains unchanged; high school graduates progress to higher diplomas at native white rates	**0.35**	0.80
3a–3c. Men in each group complete high school at the native white rate		
Scenario 3a. NONE of the new high school graduates progress to higher diplomas	**0.27**	0.08
Scenario 3b. HALF of the new high school graduates progress to higher diplomas at the rates prevalent in their own group today	**0.37**	0.12
Scenario 3c. ALL of the new high school graduates progress to higher diplomas at the rates prevalent in their group today	**0.48**	0.16

Note: Scenario 1 is the total amount of earnings ethnic men would gain if all educational differences from native whites were erased. It is the dollar value predicted from regression results described in text, note 2. The other scenarios express, as proportions of this total, the amount the ethnic men would gain from erasing specific (more limited) educational differences from native whites. See also Perlmann (2005, 104, 108).
Source: IPUMS dataset for census (2000).

whites do. Affecting high school graduation rates implies a different set of societal (and indeed familial) policies than affecting educational persistence *among* high school graduates. I want to insist upon the point that the vulnerability at the lower level, failure to complete high school, has crucial implications; table 12.1 explores this issue.[2] Specifically, the table deals only with the ethnic earnings differences associated with education, and it presents estimates of how much of the education-related earnings difference would disappear given various scenarios for change in Mexican second generation educational attainment.

In scenario #1, both minority groups reach the native white education profile; in that case, both groups would gain the entire dollar amount that they currently lose due to educational differences from native whites, and the proportions shown are 1.00 for all three groups. The other two scenarios explore how much of this gap is related to high school and college completion rates. The assumption in scenario #2 is that current high school graduates in each group would go on to college at the same rate as current native white high school graduates. In this case the Mexican group would gain 35 percent of the entire amount that the group loses as a result of educational differences from native whites.

In scenario #3, the assumption is that high school graduation rates for each group would rise to the current rate for native whites. The three variants of scenario #3 model three different possibilities for the later educational attainment of the new high school graduates added by this scenario to the current graduates in each group. The first variant, the most pessimistic, envisions that *none* of the new high school graduates would go on to post-secondary schooling. The most optimistic variant, scenario #3c, envisions that *all* the new high school graduates will match the college attainments of the *current high school graduates in their own ethnic group*. Note that this scenario #3c still assumes that these new high school graduates will be *far less likely* to finish college than native white high school graduates; in this sense, our most optimistic scenario envisions far less than perfect equality of educational behavior. Finally, the intermediate scenario, #3b, is that *half* of the new high school graduates in each ethnic group would receive *no further schooling*, while the other half of the high school graduates would match the college educational levels of the current high school graduates in their own ethnic group.

Under the most pessimistic variant, scenario #3a, 27 percent of the total dollar amount related to the ethnic difference in education would be recouped. Thus wiping out the Mexican high school dropout problem would raise earnings only modestly less than sending all current high school graduates through college at native white rates (scenario #2, which erased 35 percent of the total gap). In the two more optimistic scenarios, with some or all new high school graduates reaping still further economic rewards from having attended college, the gain from wiping out the Mexican high school dropout problem *exceeds the gain* from scenario #2, in which all current Mexican high school graduates proceed through college at the native white rate—slightly in scenario 3b and substantially in scenario 3c.

In sum, the claim that upward mobility is possible only for minority group members who complete college seems futuristic at best; above all, the claim should not divert us from noting the need for radical improvement in Mexican high school graduation rates today. The point is *not* that graduating from high school is nearly as rewarding as graduating from college; of course, college graduation produces the far greater payoff. And I am not arguing against efforts to boost the rate of Mexican American college attendance. The point is rather that *despite* the much higher returns to college compared with high school completion, the lower—but not negligible—returns to high school graduation matter for the individual. Moreover, there are *a great many* among the Mexican second generation who are not passing the lower branch point. Consequently, when our perspective shifts from the individual to his ethnic group, the leverage of changes at the lower branch point can be as great or greater for this group than even a quite radical change at the higher branch point.[3]

Outcomes are very different for the native blacks; for them the economic problem tied to educational differentials lies much less at the lower levels of educational attainment and much more in college completion. Even under the assumptions of scenario #3c, blacks would gain far less than they would from scenario #2.

The Concern About Focusing on High School Graduation Rates

As I mentioned, a colleague was disturbed by my emphasis on high school rather than college graduation rates. He pointed out that minority group members often regard such an emphasis as patronizing—as implying that the group really has no business aspiring to college graduation, that group members are "destined" for low-grade work—and that such an implication will turn out to be a self-limiting strategy for the group. My colleague gave the example of how, on a trip to Los Angeles, President George W. Bush had urged educational attainment upon a Mexican American audience by telling them that he was giving more support to a vocational high school program for them. Understandably, the speech infuriated his audience.

Reflections: General

I want to make a more explicit, if simple-minded, point about that interchange. Any discussion of the scope of educational reform can fall into this line of argument. The kind of debate that often ensues should be very familiar to anyone who spends time in educational institutions and thinks about social inequality. On the one hand is the person who criticizes an effort to raise a group's educational performance to levels less than equal to those of the most favored ethnic group or social class. It is self-defeating, so this argument goes, and recreates the inequalities in the class structure. On the other side is the person who criticizes naïve egalitarianism in favor of a down-in-the-trenches willingness to really bring some improvement, even if it is only a modest improvement. These opposing approaches have surfaced in many discussions of educational reform over the course of modern American history: in the contrast between W.E.B. Du Bois and Booker T. Washington over industrial versus collegiate education for African Americans (Du Bois 1903 ch. 3 and passim), in discussions of vocational programs in the northern cities during the same years (Counts 1922, Perlmann 1985), or in discussions about the relative merits today of stimulating community college attendance or the attendance of the select few at elite universities.

The only observation I want to make about this tired contrast of approaches is that it is impossible to call for improving one level of educational attainment—high school or college graduation, for example—and *not* be subject to the criticism that you are ignoring the other level. That is, one criticism or its opposite can *always* be made—against reform at the higher or at the lower level of schooling. I mention this simply as a consciousness-raising exercise. That is, if one is not conscious of both these arguments, and how they can pop up, then whichever critique is made can carry the day simply by invoking the guilt of anyone who calls for reform at either a lower or higher educational level. Of course, one or the other critique will carry more force in a particular context. But that is just my point: in this matter, context is all. Either critique in the abstract—stripped of any reference to particular rates of graduation and to the impact of reform on those rates—should not distract us.

Increasing College Graduation and an "Underclass" for "Segmented Assimilation": A Simple Model

There is, however, another reason to address the Mexican American high school dropout rate directly, and not to focus merely on college graduation. It is simply not the case that increasing the college graduation rate will necessarily take care of the problem of high school graduation rates (as my colleague also argued). It is true that, if the high school graduation rate is increased, some proportion of the new high school graduates will consider staying in school still longer, and some of these, in turn, will eventually complete college. It may also be true that increasing college graduation rates will set an example for actual and potential high school dropouts, and thereby increase high school graduation rates in the minority group. But note that these expectations hinge on very different causal mechanisms. To argue that the high school grad will consider college more readily than the high school dropout is an argument about the closeness of the individual to the goal in terms of years of schooling completed. But to argue that the high school dropout will consider completing high school, or that a potential dropout will remain in school, because the proportion completing college has risen is an argument about either role models or the fear of being in a shrinking low-skill pool. The strength of these latter dynamics, in turn, will depend on the relative sizes of the three populations at issue: high school dropouts, high school grads, and college grads. And among Mexican Americans, it is altogether too risky to assume that addressing college graduation rates will appreciably pull along high school graduation rates.

I mentioned at the outset that at least a quarter of Mexican second generation young men fail to complete high school, while the fraction among non-Mexican native whites is fewer than a tenth. College graduation rates for the two groups are 9 percent and 30 percent, respectively (see table 12.2, row 3, actual patterns). These college graduation rates can also be restated in terms of the proportion of *high school graduates* in each group who continue on to graduate from a four-year college—12 percent and 33 percent respectively (row 5, actual patterns). We come now to the point of this exercise. Suppose that educational reform efforts targeted at getting Mexican high school graduates through college intensify—what would be the best outcome we could expect? It would be unrealistically optimistic to expect that as many Mexican American high school graduates would complete college as is the case among native white high school graduates—since the former no doubt differ from the latter in social class and other advantages. But suppose that we *could* meet that unrealistic expectation. The resulting massive rise in Mexican American college graduation would still involve only 21 percent more of Mexican American high school graduates than it does now (up to the white rate of 33 percent from the current Mexican rate of 12 percent—row 5, hypothetical change). And the net effect in terms of all Mexican American men in the age cohort would be to raise college graduation from 9 percent to 25 percent (row 3, columns a and c). This would, of course, be a staggering improvement. But my point here is that such an improvement would touch the lives of only 16 percent of the Mexican age cohort (25–9 percent). And that change, affecting a small fraction of the age cohort, could easily be accomplished *without having any impact* on the considerable high school dropout population.

All this calls to mind the fact that a serious problem of urban inner-city poverty can develop in a substantial proportion of a group without that problem necessarily ensnaring the majority of the group. We have come to doubt the adequacy of the term "underclass," but back when it seemed like it had a clear meaning, it might have been usefully invoked here. Similarly, I at least have come to doubt the adequacy of segmented assimilation theory. But that theory, like the term underclass, might be helpful here. After all, segmented assimilation theory in its most careful formulations asks

Table 12.2 Male educational attainment among second generation Mexicans and native whites: actual and hypothetical

Educational attainments	Actual patterns		Hypothetical change	
			rate b5 rises to rate a5	impact: col. b vs. col. c
	Native white % a	Second gen. Mexican % b	Second gen. Mexican % c	Second gen. Mexican % d
All in the age cohort:				
1. High school dropout	9	25	25	0
2. High school graduate, LT BA	61	66	50	−16
3. BA (4 year college graduate or more)	30	9	25	+16
Total	100	100	100	–
Among high school graduates only (rows 2 +3 above):				
4. Do not complete a BA (row 2)	67	88	67	−21
5. Complete a BA (row 3)	33	*12*	*33*	+21
Total (all high school graduates)	100	100	100	–

Note:
rows 4 and 5. Proportion of college graduates expressed as a proportion of high school graduates
col. a. Mexican second generation: estimated from Mexican born who arrived in the United States by age three (and CPS data);
col. b. Native white: non-Mexican native whites
col. c. Hypothetical change: Mexican second gen. high school grads complete a BA at native white rate (change in row 5).
Source for actual data: IPUMS dataset for census (2000); see Perlmann, *Italians Then, Mexicans Now*, ch. 3.

us to focus on a particular fraction of a group's young people, those who have been left behind in today's dynamics of immigrant and ethnic upward mobility. It would be a wonderful thing if the small proportion of Mexican Americans who today graduate from a four-year college were to double or triple. But those results could still leave untouched the huge numbers who today do not complete high school (an underclass in making)—unless we pay specific attention to the problem of Mexican American high school graduation rates.

Notes

1. Because of the specific form of the census questions, I actually used a proxy for the second generation, namely, those who had been brought to the United States before their third birthday, but for simplicity here I speak of the second generation. The Current Population Survey, which allowed for the study of the actual second generation, confirmed the conclusions drawn from the census proxy. The discussion in this section is drawn from Perlmann (2005), 78–82 and 106–109.
2. Results are from a model in which earnings are regressed on ethnic dummies, individual age, geographic location, and educational categories. I then calculated what the mean earnings would be for each ethnic group of interest under given various proportions of the group reaching educational levels of interest. Instead, I could have used separate regression models for each group, thereby obtaining returns for schooling that include all the higher-order interactions of ethnicity, schooling, and earnings. But since the returns to schooling did not differ in a statistically significant way between native whites and the Mexican second generation, this strategy would have involved its own minor distortions.
3. Higher-order interaction terms for ethnicity, schooling, and earnings are not statistically significant despite large samples.

13

Increasing Concrete Knowledge and Community Capacity: How CUNY and other Institutions Can Help Reshape Mexican Educational Futures in New York

Robert Smith

Introduction

This chapter will present a combined history, analysis, and prescription for interventions aimed at helping create alternative Mexican educational futures in New York City. First, I offer a brief overview of Mexican migration to and settlement in New York and discuss why the Mexican population in New York both confronts serious challenges and possesses valuable resources that will help in addressing these challenges. Next, I outline a strategy for positive intervention that involves the Mexican community and a variety of New York institutions, including the City University of New York (CUNY), New York City public schools, the Mexican Consulate, private foundations, and Mexican community groups. The emphasis is purposefully on pragmatic analysis and intervention.

To telegraph the message of the chapter: Mexicans in New York present some of the most distressing educational statistics of any group in the city, and they are among its fastest growing groups. They also present a paradox: the community is internally very densely organized in small scale organizations, but it has relatively few organizations that articulate with larger scale American institutions. CUNY has begun to take significant steps to create such institutional linkages: through its Mexican Initiative and the Chancellor's Committee on the Mexican Community, and via its partnership with other organizations, including the Mexican Consulate. This chapter lays out the argument for creating and expanding such an initiative, which could be used as a model for other new immigrant groups. Specifically, I lay out a plan for increasing (1) the Mexican community's concrete knowledge of how the educational system works in New York,[1] and (2) the Mexican community's collective capacity to promote its youth's educational success, both internally and through strategic collaboration with larger American institutions in New York City. Increasing community capacity to relate to larger institutions is a wise investment because so much public work is now done through public private partnerships or alliances.

Theoretical Considerations

While the primary purpose of this chapter is diagnosing a problem and outlining a practical plan for intervention, the analysis is also theoretically relevant. In the 1920s, Robert Park (1922) bucked the contemporary wisdom—which held that immigrant organizations represented impediments to Americanization and hence should be discouraged—to argue that immigrant organizations and institutions, such as the immigrant press, should not only be tolerated but encouraged and engaged by mainstream America, because they facilitated assimilation. The current wave of immigration has similarly occasioned scholarship on the structures and helpful features of immigrant organizations (Cordero-Guzman 2005, Moya 2005, Schrover and Vermeulen 2005,). These works inform my analysis, which seeks to build bridges between immigrant organizations and larger institutions such as CUNY.

I also draw on three related veins of work on nonprofits. The first documents how nonprofit organizations form a beneficial institutional bridge for youth in poor neighborhoods and have become more important, as the state has increasingly withdrawn support from social service programs (Furstenberg et al. 1999, McLaughlin et al. 2001, National Research Council and Institute of Medicine 2002, Roffman et al. 2003, Deschenes et al. 2006). Second, scholars argue that government has a positive role to play in promoting the development of immigrant organizations. Contrary to those who fear government will crowd out a Toquevillian, self-organized civil society (Joyce and Schambra 1996), Irene Bloemraad

(2005) emphasized that state policies toward immigrant organizations can have a positive effect on how they develop. Finally, Salamon and his colleagues (1999, 2002, 2006, Boris and Steurele 1999) argue that the relations between the government and nonprofits have become more complex over time, and that nonprofits thus need to develop appropriately sophisticated tools to negotiate these relationships.

In the plan I outline, CUNY, the Mexican Consulate, and Mexican and mainstream organizations all work together to improve the educational futures of Mexican youth. This plan will help develop the capacity of Mexican organizations and build institutional bridges between them and public institutions—here, CUNY and the New York City public schools. New York is a particularly positive environment in which to promote such community development, given its long history and plethora of institutions focused on positively incorporating immigrants (Mollenkopf 1999, Cordero-Guzman 2005). CUNY's wide geographical reach also matters here, because the Mexican community is dispersed throughout the city.

Mexican Immigration and Education in New York: Challenges to Educational Success

Mexicans are a key community in New York City's educational future. They are the fastest growing major population group in the city, with about 500,000 people of Mexican origin in New York in 2010, up from about 40,000 in 1980. Mexicans have surpassed Dominicans as the group accounting for the largest number of births in the city (Salvo 2005). On details of the growing Mexican immigration to the east coast, see Zuniga and Hernandez (2005), and Smith (2006).

Many Mexicans in New York in the 1990s developed an anguished relationship to the New York City public school system. Their relatively sudden emergence as a visible ethnic group occasioned a sometimes hostile reaction among some natives in their schools and neighborhoods, often Puerto Ricans. Mexicans also have low levels of parental education—about six years—and the highest rate of 16–19-year-olds not graduated from high school and not in school, 47 percent, versus 22 percent for Puerto Ricans, 18 percent for African Americans, and 7 percent for whites, according to the census. Between age 14 and age 19, there is a 69 percent drop in school enrollment, from 95 percent to 26 percent for boys; the drop for girls is lightly less, from 96 percent to 31 percent (Rivera-Batiz 2004). These are sobering numbers, threatening a pessimistic future.

These high nonattendance rates comprise both those who have dropped out and a large number of those who have never "dropped in." Many

Mexican youth are "teen migrants" (Smith 2006) who may not have entered school in New York, or dropped out to work after only a short time in the system. The end result is an educational hemorrhage, but such complexity suggests different kinds of interventions to reach "real" drop outs, versus teens who were never in school. Different kinds of "Mexican" students—U.S.-born citizens, legal residents, teen migrants, or the undocumented—present different challenges to the school system (see Suárez-Orozco and Suárez-Orozco 1995, Suárez-Orozco et al. 2006).

Mexicans have correspondingly low rates of college enrollment. Only 1.6 percent of CUNY applicants in 2005—fewer than 800—were Mexican, out of an 18–25-year-old Mexican population I estimate to be at least 50,000.[2] While most are not high school graduates, and hence could not directly apply to CUNY, the relatively small numbers applying to CUNY indicate the gravity of the situation and the seriousness with which CUNY must approach this issue.

In addition to these harsh statistics, the Mexican community as a whole suffers from a lack of knowledge about the public and higher educational systems in New York and a lack of social infrastructure supporting education. Parents are fervent in their belief in education for their children, but usually lack the knowledge of *how* to help their children. Finally, many Mexican parents and youth believe that college is too expensive for them to afford or requires full time attendance. Yet CUNY has varied programs, and full time tuition at community colleges is only $1550 per semester. While this is a lot for these youth, it is not impossible, and is not consistent with the notion that only *los ricos* (the rich) can attend college.

The problem of social infrastructure related to education is multifaceted. Part of the problem is a simple lack of knowledge, which can be combated through dissemination of information. But another part of the problem is that most Mexican community groups are not well enough funded to mount effective educational intervention programs or alliances with schools. This shortage of funding has many causes, including the lack of articulation between different parts of the larger Mexican community. This point can be understood via comparison with the Chinese community. According to a long-term study of mobility among children of migrants (Kasinitz et al. 2008), the Chinese as a group have similarly low levels of income and education as do the Mexicans, but the Chinese elite have promoted the Chinese educational achievement through their investments in community organizations. These wealthier Chinese value the honor they are accorded as community leaders, and support community organizations—the myriad after-school, academic, and Chinese language and summer programs that help boost Chinese academic achievement. Programs of these kinds, especially summer ones, account for much of the achievement gap between whites and blacks (Smith and Brewer n.d.), and we can easily theorize that they can help explain the differential achievement of Chinese and Mexicans. Yet aside from a few migrants turned entrepreneurs, most of the elite in New York is unengaged with rest of the Mexican community. My hope is that a stronger engagement between the Mexican elite (especially those who work in financial, corporate and legal worlds) and grassroots Mexican organizations can be developed, and there are positive signs in this direction.

A key factor framing Mexicans' relationship to the school system is the undocumented status of many parents and students. The undocumented population is growing steadily in the United States and New York. Passel (2005) estimates there are over ten million undocumented persons in the United States; 57 percent are Mexican. Each year, about 60,000 undocumented students graduate from American high schools, with 10 percent of these in New York (Passel, Capps and Fix 2004). A study done at CUNY showed that 76 percent of undocumented students surveyed had graduated from high school in New York, indicating longer-term residency (Dozier 2001). In a recent study of access to medical care in New York, McNees, Suilc, Flores, and Smith (2003) found that 91 percent of a convenience sample of 561 Mexicans—contacted in public places—who had been in the United States for up to ten years were still undocumented, and that this only fell to 83 percent after 15 years. All this suggests that increasing numbers of undocumented immigrants are moving from adolescence into early adulthood and parenthood while still undocumented, greatly affecting the developmental contexts they can construct for their own children.

This undocumented status has several key impacts. First, it discourages parental participation in the schools. Second, and perniciously, undocumented status inhibits the aspirations of many Mexican youth, who think being undocumented precludes them from attending college. In a convenience sample of more than 500 Mexican potential CUNY students, some 90 percent were undocumented, and 90 percent of those believed one could not attend college if undocumented.[3] This is not the case—indeed it is even possible under certain circumstances to get scholarship money at CUNY. But if students believe it is so, this belief, and not CUNY policy, will determine their actions. However, just as this erroneous belief has cumulative negative effects, so can the dissemination of correct information and the advocacy of current Mexican college students become a cause of higher rates of attendance.

Some students, and their parents, who do know they can apply to college, still ask quite reasonably what will happen when they finish? This is the crucial but often overlooked issue when we talk about undocumented students and higher education. There is no guarantee that getting one's degree will help in getting legal status, and many CUNY graduates cannot find jobs commensurate with their educational status because of their legal status. Yet, what impresses me most is that so many students do forge ahead despite the uncertainty.

Mexicans also face disadvantages in New York because they are geographically dispersed into some six "little Mexicos" in New York City. This dispersal dissipates the focus of attention of the problem away from one political district, so public officials are less likely to take notice, and no one politician will be seen to "own" the problem. However, the first large cohort of US born Mexican Americans in New York is coming of age. Fostering civic consciousness and voter registration among them is one positive intervention CUNY can do to promote Mexican well-being in the city.

These problems are compounded by the school choice system in the city's high schools (Smith 2002, Mollenkopf et al. 2005). Explained briefly, high schools admissions are not based solely on where you live, as it is in most places. De facto, there are three tiers of schools—elite, partly selective and zoned schools. Elite schools are competitive, and only one person of our whole sample (Smith 2002) attended one. Partly selective schools admit some students by their being in the catchment area, but because of strong academic reputations, attract many from outside it. Admission is at the discretion of the school. Upwardly mobile informants were likely to go to one of these schools. Finally, the zoned school is the one that must accept all comers in the catchment area who have graduated from middle school. In an earlier study (Smith 2002) school type strongly affected our informants' experience of school. Zoned schools tended to have harsher inter-ethnic/racial relations, worse relations between students and teachers, more gang activity—especially among Mexicans in those schools—and lower graduation rates. Boys in our sample were more likely to go to zoned schools and end up experiencing their Mexicanness as a racial, oppositional identity in line with the predictions of segmented assimilation theory, while the girls, especially in partly selective schools, were more likely to experience it as one identity among others, in a more cosmopolitan pan-minority or pluralistic setting that also included white students.

In sum, the Mexican community faces several serious challenges to its educational future.

Community Resources

The Mexican community in New York faces many challenges, but also has significant resources to confront them. While the real per capita income of

Mexicans in New York has decreased significantly over the last 20 years—from $17,495 in 1980, to $13,537 in 1990 to $15, 631 in 2000—and is among the lowest of any group, they have higher rates of *household* income than other groups because they have more wage earners per household, and they have a higher percentage of households in the middle class (making over $50,000 per year) than Dominicans or Puerto Ricans (Salvo 2005, Bergad n.d., Population Division, City Planning Department 2004). This pattern is part of a larger strategy of collective upward mobility that some immigrants in New York have pursued (Mollenkopf et al. 2005). Hence, in my current research, I see youth in their mid and late twenties living with their parents, even after marriage sometimes, in an attempt to save money and give themselves a better platform from which to launch their own lives at a later date.

A second and underutilized resource is that the Mexican community is internally very densely organized. There are in New York and the region a myriad of home town associations, soccer clubs and leagues, religious organizations, and community organizations of various types. But these organizations are not well linked up to American institutions (Breton 1964). In particular, they do not have strong relations with the New York City public school system. Nor has the Mexican community formed a series of support organizations oriented towards schooling, although a few exist and do impressive work, such as the Adelante Alliance's Distance Learning Program. This is not a criticism of these organizations or leaders, who address pressing problems such as legal status or discrimination. Rather, in line with the main purpose of this book, I seek to identify a constellation of factors producing a negative outcome on education, and to change them.

This lack of articulation with American institutions and orientation is curious, because when you ask Mexican parents what matters most, they say, without hesitation, "education." So our collective task here is to figure out how to link up the tremendous desire of Mexican immigrants to help their children with Mexican and with larger American institutions, foundations, and community organizations. This is where the Mexican Initiative and the Chancellor's Committee, Baruch College, and Mexican Mentorship Project come in.

Increasing Concrete Knowledge and Community Capacity: CUNY's and Community Organizations Work Together

The CUNY Mexican Initiative and Chancellor's Committee, Baruch's Mexican Consulate-Mexican Communities Program, and the Mexican Mentorship Project work closely together seeking to promote educational achievement among Mexicans in New York. The Mexican Initiative and Chancellor's Committee and Mentorship Project are based in part on the foregoing analysis and are still works in progress.

The Mexican Mentorship Project

The Mexican Mentorship Project is a joint venture between the Mexican Educational Foundation of New York (MexEd) and the Mexican American Students Alliance (MASA). The project's mentoring work does not need to wait for large-scale changes in society to help Mexican youth. Rather, we seek feasible sites of intervention, identified through our community involvement, our research and our life experience, that will enable us to make a positive difference in the lives of the students with which we work.

We believe that such local level intervention can have very positive consequences, which can be replicated. We seek first to create relationships between our mentees and our mentors, who are educationally and professionally successful Mexican Americans, Mexicans, and others in New York. Mentors help with academics, applying to college, and in other, more informal ways. We also seek to strengthen the relationship between the parents and the school, and to give both parents and students information related to college applications, making the process less mysterious and forbidding. Our mentors also advise parents on dealing with the school system, and will help alert the school to issues in the community it may not know about but should. Concretely, MexEd and MASA have run after-school programs and a mentoring program for students of high school age, and some younger ones, and offered English lessons for their parents; they have done outreach for CUNY (discussed more later); and aid students looking for scholarships.

The mentoring relationship we have established follows best practices laid out in the work of Jean Rhodes (2002, Rhodes and DuBois 2006). We seek long-term commitments to our mentees, usually a year. A way to understand these mentoring relationships is through the concept of social capital, which refers to relationships that give one access to resources (financial, informational, or simple "know how") that facilitate success (Portes and Rumbaut 2001). We hope to expand the social capital we help our students generate through a future professional mentoring program and other programs.

The Mentorship Project has faced several obstacles. The most critical is financial, and another one has been that the demand for mentors outstrips the supply. We have recently begun relationships with local universities, which are offering us students; the students, in turn, get get supervision for service learning. We are also seeking to work with other

Mexican organizations, with CUNY, and with the New York City public school system more widely, but these efforts will require increased funding.

CUNY's Mexican Initiative and the Chancellor's Committee on the Mexican Community

CUNY's Mexican Initiative is a broad-based effort that has emerged over several years, and has recently (in 2007) been more formally established as the Chancellor's Committee on the Mexican Community. In this section, I relate this initiative's history thus far, and then briefly discuss the ambitious plans of the Chancellor's Committee. The Mexican Initiative has been propelled by the commitment of CUNY's chancellor, Matt Goldstein, and the vice chancellor, Jay Hershenson, as well as support from the Offices of Admissions and of Continuing Education, and by Dean David Birdsell of Baruch College's School of Public Affairs.[4]

A primary document cementing the initiative is the Memorandum of Agreement signed in 2005 by CUNY chancellor Goldstein and then Mexican consul Arturo Sarukhan. This memorandum recognized the growth of the Mexican community in New York, as well as its educational distress, and laid out concrete areas of cooperation for the two institutions to address the issues. The first plank of this initiative is a multifaceted information dissemination campaign directed particularly at the Mexican community.

A second plank is targeted college outreach to specific high schools that have the largest numbers of Mexican and Mexican American students. These will not be one-time visits, but outreach followed by technical support in filling out applications. The idea is to have CUNY admissions counselors go several times to the same school, first to give out the information, and second to follow up with issues that may have emerged with particular students' applications.

A third plank is the CUNY Ambassadors Program, which began in 2006 and trained more than 40 members of Mexican community organizations in how to help Mexican students, including the undocumented, to apply to CUNY. The Ambassadors program seeks to ensure that community leaders know how to address these issues and have CUNY staff whom they know personally to call with questions and referrals. The Ambassadors program both increases concrete knowledge in the community and helps create links with one larger American institution.

A fourth plank I would like to add to the initiative is differentiated outreach and organization. One way to do this is foster greater parental involvement in the New York City public schools. A second way would be to focus on Mexicans and Mexican Americans who are currently in school, using the advantages of having them in school to help them transition to college. A third way would be to focus on those who have not or are not currently in school, the "never dropped ins." They have been extremely responsive in MASA or MexEd outreach events, especially those advertized through Spanish language media. These students can benefit from Graduate Equivalency Diploma (GED) programs, and information about community college. For others, information about basic Spanish literacy is in order. It would be possible to reach a large percentage of Mexicans by focusing on a relatively small number of schools in which Mexicans are concentrated. These outreach efforts would also do well to systematically include active community organizations, as well as to involve local churches attended by the immigrants and their children. Several churches have been very supportive of these efforts, offering free space for outreach events and even ongoing programs.

A final, and very promising, evolution in CUNY's relationship with the Mexican Community is the formation of the Chancellor's Committee on Mexicans, Education and CUNY in 2007. Its subcommittees include those for dropout prevention, scholarships, research, relations with the New York City public school system, and others. This Committee seeks to more fully commit CUNY, with its 23 colleges and over 200,000 students, to help the Mexican community develop its educational potential. There is also discussion of the need to foster early childhood education and use of pre-K among Mexican immigrants. All these efforts are attempts to intervene with the first large cohort of children of Mexican immigrants, to get in "on the ground floor" of an unfolding social issue, and to make a positive difference.

Baruch's Mexican Consulate-Mexican Communities Program

Baruch College's School of Public Affairs (SPA) has also sought to enhance the Mexican community's ability to help itself. Each year, beginning in 2006, SPA, the Mexican Consulate and the Vice Chancellor's Office have held an eight-week Saturday session Emerging Leaders in the Mexican Community course. These programs train 20–40 leaders in the Mexican community groups in New York, New Jersey and Connecticut each year in various aspects of how to run a nonprofit organization, apply for grants, build a board of directors, and other matters. Know-how of this kind is sorely needed in the Mexican community. The Mexican Consulate-Mexican Communities Program was modeled on SPA's long-standing Emerging Leaders Program. The program has benefitted from the active support of established nonprofits like the United Jewish Committee, and the mentorship of established foundation and nonprofit leaders, including some from the Hispanic Federation. Having older nonprofits share their experience with Mexican newcomers will facilitate their development of social capital and organizational capacity, as noted in the work of various scholars (McLaughlin 2001, Portes and Rumbaut 2001, Salamon 2006, Kasinitz [in this volume]).

Finally, an unanticipated positive outcome to the Baruch classes is that community leaders have met other leaders they did not know before. This has fostered sharing of approaches to common problems, and has facilitated a a consensus about the need to do something collectively about education for their children. There has been fruitful discussion about forming some kind of Congress of Mexican Community Organizations for Education (COMEXCOE) and working with CUNY, SUNY, and public universities in New Jersey, and the public education systems in New York and New Jersey, to promote the education of Mexican youth.

Conclusion

In conclusion, I wish to address the collective action and financial problems confronting Mexican organizations in New York. Part of the problem moving forward is that each small organization is doing its own work and scrambling for funding or other support, sometimes competing with each other (Franco 2007). The issue is that there is no real umbrella organization that can coordinate larger action. What is needed is a coordinated strategy that would involve CUNY, the New York City public school system, selected New York City public officials, the Mexican Consulate, and a variety of Mexican organizations. Existing organizations with greater capacity, such as Associacion Tepeyac, could play key roles in promoting this strategy. What I envision is a large mobilization project, with a scholarly/evaluation component, that would span three to five years and coordinate action among all these players. The goals would be to promote school success and college attendance; promote Mexican community capacity, especially in nonprofit organizations; build stronger alliances and political links between Mexican organizations and New York City institutions, and the Mexican Consulate; and make these Mexican community organizations more financially sustainable, enabling them to graduate to the next level in their nonprofit development. What I envision would require sustained, medium-term funding, which would be justified by how urgent the current situation in the Mexican community is and how propitious the moment is for intervention. Having sustained mid-term support from important leaders in the New York City and State government would also be extremely valuable in promoting such an initiative. Developing a larger entity by which the Mexican community could act collectively in educational issues, such as COMEXCOE, would be a useful step. What is needed is support to channel all this desire to help into real programs and increased community knowledge and organizational capacity.

Mexicans' current scenario in New York is an educational tragedy happening in slow motion. Left unaddressed, it portends a harsh future for Mexican youth and their children in New York. Yet this future is not written in stone. It is one of several possible futures, and we as a society can act to alter that future, to improve the chances these students will have and the choices they will make. It has been a great source of satisfaction and pride to be involved in CUNY's efforts to intervene while the first cohorts of Mexican-born youth are coming of age in New York, to proactively address the educational needs of a community that could become New York's largest ethnic group. A better future can be written.

Notes

1. "Concrete knowledge" is a phrase coined by Sandra Lara to describe what the Mexican community lacked, that is, understanding of specific steps to be taken, even to apply to college. Sandra (now Dr. Lara-Cinisomo) was a cofounder of the Mexican Educational Foundation, and was a researcher for me in the mid-1990s.Thanks to Blanca Ibanez for all her help with these programs and my other work.
2. I estimate that there were about 500,000 Mexicans in New York City in 2007, most of whom were under age 30. I conservatively estimate there are at least 50,000 people, or 10 percent of the total, in the 18–25-year-old age range. The actual number is likely to be higher. I am assuming here that the number who apply to CUNY reflect a significant part of all applicants to college, on the basis of my research, and given the financial realities most Mexican families confront. The 1.6 percent figure comes from CUNY's Institutional Research Office; my thanks to Dr. David Crook for supplying data on Mexican enrollment.
3. These are eyeballing estimates from a perusal of the data, which have not yet been fully analyzed.
4. The initiative has committed work of Richard Alvarez and his staff, of the Office of Admissions, John Mogilescu and his staff in the Office of Continuing Education, and Baruch College School of Public Affairs dean David Birdsell and Executive Programs head Anne Reucker and their staffs.

Part V

Institutions Providing Aid to Young Immigrants and Refugees

A major goal of this endeavor has been to bring together scholarly and practical perspectives on the topic of young immigrants and refugees. Whereas, in the preceding chapters, leading scholars probed various aspects of this issue, representatives of major aid organizations and of the U.S. government discuss their institutions' activities in aid of young immigrants and refugees in Part V.

Scholars tend to take a long view on immigration, focusing on long-term causes and consequences. Many of these causes and consequences are deeply anchored in enduring societal structures and thus not directly amenable to quick change. Organizations tend to have a short-term approach. They focus on immediate action and aid, and their measures of institutional efficacy and success also tend to be short term.

The chapter by Robert J. Carey and Jane S. Kim, submitted on behalf of the International Rescue Committee (IRC), presents in detail the activities and lessons learned by this organization from them. As a major challenge, the authors identify the wide-ranging diversity (and consequently the diversity of needs) of the new waves of refugees/immigrants from all parts of the globe.

The chapter by Julianne Duncan, on behalf of the United States Conference of Catholic Bishops, focuses on the question of what would be meant by success in helping young refugees and other migrants. Duncan emphasizes that the root of successful aid is to instill a sense of belonging and of purpose in life in the young person.

The contribution by Annie Wilson, of the Lutheran Immigration and Refugee Service (LIRS), centers on the role of nonprofit organizations among refugees and other asylum seekers, or migrants. The chapter then covers the guiding principles of the nonprofits in terms of real action, specifically by the

LIRS. The framework for identifiable success includes, at the broadest level, "protection, stability, and integration."

Complementing the chapters by representatives of nongovernmental aid organizations, a perspective from inside the federal government is presented by Ken Tota, of the U.S. Office of Refugee Resettlement (ORR). Within the federal government, ORR is charged, among other duties, with the care of unaccompanied alien children arriving in the United States. The chapter provides a framework for the care of these children that spells out the fundamentals that need to be in place for services to be provided and to be efficacious.

14

Tapping the Potential of Refugee Youth

Robert J. Carey and Jane S. Kim
International Rescue Committee

Introduction

Long gone is the era of conventional warfare—as the predominant casualties of war today, children disproportionately and profoundly bear the brunt of conflicts. Since 1987, an estimated 2 million children have been killed in armed conflict, as either noncombatants or child soldiers. One million have been orphaned. More than 6 million children have been seriously injured or disabled. Countless more children suffer disease, sexual exploitation and abuse, forced labor, malnutrition, or psychological trauma as a result of violence they have witnessed, committed, or endured.[1]

Roughly 30 armed conflicts raging worldwide have forced the displacement of some 30 million people, half of them children. With entire communities, schools, homes, and families being destroyed, war-affected and displaced children are stripped of their childhood and, at times, their humanity.[2] Refugee and displaced youth in particular are more vulnerable to specific risks, including forced recruitment in armed conflicts, lack of education and vocational training, sexual abuse and exploitation, HIV/AIDS and other health threats, human trafficking, and economic exploitation.

Familiar is the refrain that "children hold the key to our future." And increasingly, there is recognition of refugee youth as extremely resourceful and exceptionally resilient, with boundless potential as positive contributors to our society—attributes all the more critical to successfully negotiate the complex and broad array of challenges facing them in the United States. Yet, this is not fully acknowledged or addressed by federal and state legislation, educational policy and practice, specialized youth programming nationwide, or comprehensive research; nor are refugee youth viewed as a target population as reflected in current funding priorities and levels. Regrettably, refugee youth in the United States continue to remain underserved.

U.S. Refugee Resettlement Program

For more than two decades, the United States has upheld a rich tradition of offering refuge to those who have suffered past, or fear future, persecution. With the passage of the Refugee Act in 1980,[3] the U.S. government signaled an enduring commitment to assist refugees[4] of special humanitarian concern by authorizing federal funding to formalize the U.S. Refugee Admissions Program. In response to the annual flow of incoming refugees, the U.S. Department of Health & Human Services Office of Refugee Resettlement has been charged with administering refugee benefits and services, including cash and medical assistance, employment preparation and job placement, skills training, English language training, and other support services.

- Since 1975, the United States has resettled more than 2 million refugees, roughly half of them children.
- In FY06, approximately 47 percent of all refugees—roughly 19,500—admitted to the United States were below the age of 21.[5]

With the fall of Saigon in April 1975, the United States resettled hundreds of thousands of Indochinese refugees with temporary funding. Since 1975, the United States has resettled 2.4 million refugees, with nearly 77 percent being either Indochinese or citizens of the former Soviet Union. While the United States has demonstrated particularly strong humanitarian and foreign policy interests in these two groups during the past three decades, in more recent years, there has been a palpable shift toward focusing on wide-ranging refugee groups originating from other regions of the world.

In fact, the U.S. Refugee Admissions Program (USRP) is more diverse than ever. Refugee arrival patterns have shifted dramatically during the last decade in response to changing refugee circumstances. An unprecedented number of refugees are now arriving from the varied regions of Africa, East Asia, and the Near East. During fiscal year 2006, USRP refugee applicants represented some 60 nationalities in over 50 overseas locations.[6]

And for the first time in history, the United States officially opened its doors to refugees from North Korea. Further, the United States has directed greater attention to refugees in protracted and stagnant situations, as well as stateless populations.

The sheer diversity of incoming refugee groups requires more refined and continuous planning. Resettlement challenges typically run the gamut: the dynamic and fluid nature of refugee resettlement; growing linguistic diversity; broad-ranging employment and educational histories; distinct and divergent refugee experiences; gender, cultural, and religious considerations; psychosocial concerns; job placement and economic self-sufficiency; intergenerational conflicts; educational barriers; considerable variation among refugee groups vis-à-vis specific age categories (i.e., school aged, working, and retired); increasingly fewer close family members in the United States for resettled refugees to rely on; and the shortage of available affordable housing in urban areas.

While newly arrived refugees are entitled to benefits and services—cash assistance, health benefits and services, employment assistance, career development and training opportunities, and a wide variety of social services—eligibility is generally specific, limited, and time bound. And despite compelling data revealing a school-aged population that invariably comprises a sizable percentage of overall arrivals each year, youth and children's programs and services are discretionary, and all too scarce to meet the tremendous needs of refugee youth and children across the nation.

Refugee Youth Left Behind?

Refugee youth arrive in the United States all too often having endured devastating conflict, displacement, violence, trauma, and chronic disruption of their education. A recent survey of the International Rescue Committee's (IRC) domestic resettlement network indicates that all newly and recently arrived refugee youth face multiple and wide-ranging challenges and risks at various levels and to varying degrees. Major challenges reiterated across the IRC network include academic performance, social isolation and alienation, language barriers, school neglect, and family pressures.

While refugee youth confront adversity on many different levels, academic difficulties present one of the prevailing stumbling blocks with far-reaching implications.[7] Similar to the fundamental role that education plays abroad in mitigating the effects of war-related trauma, providing structure and stability, advancing social and cognitive development, contributing to psychological and social well-being, enabling youth to regain hope and dignity, and preparing youth for constructive adulthood,[8] education is vital to the healing, healthy acculturation, and future success of refugee youth in the United States. In the last half century, education has emerged as the most likely route to status mobility.[9] For immigrant youth in particular, who are spending more time in school than ever before in U.S. history, engagement with schooling—including relationships formed and the acquisition of academic, linguistic, and cultural knowledge—will generally bring about the most profound transformation in their lives.[10]

Unfortunately, refugee youth in the United States all too often experience academic challenges. English language proficiency is one of the key factors in predicting academic achievement (see Suárez-Orozco and Gaytán, in this volume).[11] Nearly all refugee youth, however, require instruction to develop English language speaking, reading, and writing skills. Whereas elementary school children experience less pressure at school and generally learn the language quickly, older refugee youth struggle to meet demanding academic requirements in a short period of time.[12] Repeatedly, refugee youth are enrolled in public schools on the basis of age, regardless of academic background. Public school curricula and staff-to-student ratios are not designed to address the extensive needs of refugee students who perform below grade level. There are too few, if any, school personnel available to conduct linguistically and culturally appropriate assessments of the cognitive and linguistic abilities and emotional and social development of refugee students, as well as to identify and respond to refugee students with learning disabilities.[13] And whether refugees are scattered across various school districts or predominantly placed in a single district, schools by and large cannot effectively meet the needs of all refugee students. Meanwhile, refugee parents and other caregivers, also struggling with English language acquisition and overwhelmed by financial concerns and acculturation and employment demands, are frequently unable to offer appropriate support and guidance to the children.[14]

In 2001, the United States passed the No Child Left Behind (NCLB) Act aimed at reforming the educational system to increase accountability of states, school districts, and schools; offer greater choices for parents and students, particularly those attending low-performing schools; provide more flexibility for states and local educational agencies (LEA) in the use of federal education dollars; and place a stronger emphasis on reading, especially for younger children.[15] While implementation of this act varies from state to state, refugee youth are held to the same academic content and achievement standards as their American-born peers. Refugee students are generally required to take state assessment tests, without consideration of their English language ability, familiarity with standardized testing procedures, education history, or length of residence in the United States.

Understanding the impact of these tests, refugee students feel extreme stress and pressure to pass. Meanwhile, schools feel acute strain when refugee and other limited English proficient students jeopardize their position to meet standards.[16] As a result, schools with poor performance focus solely on teaching to the exam in key grades, which can deflect resources from other areas, particularly ESL/ELL (English as a Second Language/ English Language Learners) classes.[17] The emphasis on high-stakes testing not only widens the achievement gap among ethnic minorities, but also potentially delivers the hardest-hitting impact on refugee students. According to one study, newly arrived youth cannot be expected to master the intricacies of academic English in one year of study and are therefore unprepared for NCLB or other high-stakes testing, which can lead to premature disengagement from school.[18]

Refugee youth are at higher risk of failing academically than their peers. Comparable to other immigrants, refugee youth contend with the disruption of migrating to a new country, learning a new language, adjusting to an unfamiliar culture and lifestyle, discrimination and racism, and a crisis of identity in response to the cultural demands of their parents and of their new peers.[19] For older refugee youth, low academic placement with no opportunity to advance at an accelerated pace, exacerbated by difficulty navigating the U.S. educational system and pressure to work full time, can also lead to high dropout rates. Acculturation stress has been found to be considerably higher among refugees.[20] Compounded by the trauma and psychosocial pressures associated with the refugee and resettlement experience, numerous refugee groups are unlikely to possess basic levels of education or vocational skills, which hinders educational success.

These challenges unfold on multiple fronts. In 2006, an estimated 115 million children around the world were denied the right to primary education, with one-third of these children kept out of school because of the effects of conflict.[21] Not surprisingly, therefore, countless refugee children and youth arrive in the United States with limited access to schooling or vocational training in their home/host countries. This is consistent with the overall rise of ELLs across the United States—by 65 percent between 1993–1994 and 2003–2004—and the increased numbers of students with interrupted formal education (SIFE) within the general immigrant student population.[22] According to one researcher, SIFE represent "the highest of high-risk students."[23] While exact figures on the prevalence of SIFE nationwide are unavailable, New York City estimates—approximately 13.4 percent of total ELLs, or 18,900, are SIFE, and 60 percent drop out of high school[24]—are a clarion call for greater and immediate attention to this population.

Meanwhile, refugee parents, particularly from war-torn regions, have limited education and literacy skills, and tend to fall within the lower socioeconomic strata in the United States. A well-established correlation exists between poverty and educational outcomes, as well as the level of parental educational attainment and academic achievement of children. This does not augur well for certain refugee populations. Viewed together with academic disparities across the board between refugees and non-refugees, it becomes all the more critical to design targeted interventions to close this gap.

Moreover, the stress and trauma experienced by refugee youth can adversely impact their learning and academic success.[25] Traumatic memories can continue to surface. Cultural bereavement, barriers, and conflicts can arise. Survival behaviors learned in refugee camps, different conceptions of boundaries, or widely divergent social norms can clash with the dominant U.S. culture. Exacerbating this is the social isolation and alienation experienced by most refugee youth. According to one study centered on Vietnamese refugees, the perception of acceptance in a new school setting is an important factor in a refugee student's psychosocial adjustment.[26] In fact, the need for acceptance and belonging transcends any one particular ethnic group and has emerged as a common theme among all refugee youth served by the IRC network. Sadly, cultural stereotypes and prejudices channeled through acts of bullying, ridicule, and teasing greet numerous refugee students. It is all too common for African refugees to be singled out for verbal harassment by their African American peers, or for Muslim refugees to be expressly targeted for their perceived association with violence and terrorism.[27] While these antics can easily be characterized as run-of-the-mill adolescent mischief, discrimination and rejection by peers can take on heightened harmful consequences for refugee youth.

At times, teachers and school administrators are the unwitting offenders. While prejudices generally are not expressed as overt hostility, refugee students may detect subtler forms of insensitivity, ignorance, or indifference and feel marginalized by negative attitudes or exclusionary structures and practices in place. This is especially disturbing in light of evidence indicating that students' motivation and efforts in school correspond with their connection to their teachers.[28] In other cases, teachers are at a loss in knowing how to respond to refugee youth, particularly in handling disciplinary problems or potential mental health issues; at the same time, they may be unaware of the availability of community supports, where they exist. While the impact on refugee youth may not be irreparable, it certainly can affect their learning.

Family structures and community resources are also critical ingredients in positive psychosocial adjustment. Parental support, interest, and involvement are key.[29] In many instances in which refugee students thrive academically, the level of parental support is strong. Conversely, parents can present additional barriers to their children's healthy acculturation and academic success. This can stem from a cultural misunderstanding of the role and responsibility of parents vis-à-vis their child's education, or the unquestioned deference and authority conferred on teachers as the educational experts.[30] Other times, refugee parents are simply overwhelmed by economic survival concerns and the host of challenges they confront as they rebuild their lives. Intergenerational stress and conflicts can also arise for a number of different reasons. As youth generally are more adept at learning English and often find themselves acting as translators with various parties (teachers, doctors, landlords, representatives from social service agencies, etc.), parent-child roles are reversed and can create tensions.

Further, refugee youth may reject the culture of and alienate their parents as they fall sway to the competing demands and lifestyles of their new peers. As numerous refugee families initially settle in urban areas with high rates of poverty, crime and unemployment, and inferior schools, the potential for refugee youth involvement in negative social behaviors—truancy, dropout, drugs, gangs, promiscuity, violence, criminal activity—is worrisome. The refugee experience, resettlement location, and school context can make a difference with regard to academic engagement, socialization, and acculturation. Of noteworthy concern is the particular vulnerability of refugee youth from protracted conflict situations and/or refugee camp settings to engage in aggressive, violent, and high-risk behaviors especially when resettled in disadvantaged areas.[31]

The challenges facing refugee youth are distinct, complex, and pressing. After opening our doors to refugees fleeing persecution, it is extremely shortsighted and woefully unacceptable to then leave youth behind once they arrive in the United States. It is imperative that various key agencies—the U.S. Department of Health & Human Services and the U.S. Department of Education—elevate refugee youth issues to high priority and develop appropriate refugee youth policy and concrete support for action-oriented research and effective programmatic initiatives specifically for refugee youth.

Effective Youth Programming: Changing the Odds

Firmly committed to changing the odds to maximize refugee youth success, IRC's network is currently translating positive youth development principles into action. Across the country, IRC has launched a broad range of comprehensive and integrated youth programs and initiatives based on effective youth development principles while taking into account the unique needs, challenges, and strengths of refugee youth and their families.

There has been no shortage of programmatic challenges, which include engaging parents, working with schools, responding to psychosocial concerns, dealing with child abuse and neglect, conflict resolution, disciplinary issues, linguistic and cultural barriers, crisis intervention, handling domestic violence issues, and licensing/certification requirements.

Compounding this is the tremendous diversity of ethnic groups served by IRC's network and the dynamic and fluid nature of refugee resettlement. While the majority of incoming refugee groups can be anticipated to some extent, this is not always the case. As a result, this can influence a wide range of programming processes, from strategic planning and outreach to curriculum development and collaborative partnerships. Linguistically and culturally appropriate programming also demands adequate resources.

Unfortunately, there are glaring funding gaps for refugee youth programming. Further, in the midst of day-to-day program implementation, maintaining relevant knowledge of the cultural, educational, and refugee experience backgrounds of all refugee youth served—including gender and religious considerations and potential programmatic implications—can prove challenging.

Despite some of these hurdles, targeted youth programming and strategic partnerships can go a long way toward advancing the potential of refugee youth. Over the past decade, IRC's resettlement network has offered diverse types of programming for refugee youth, covering the broad areas of academic support, recreation, and life skills. While there are variances and nuances to each youth program, IRC's national network embraces an overall framework that adheres to the values and promising practices governing positive youth development; ensures comprehensive and integrated programming; and vigorously cultivates effective partnerships with refugee youth, their families, schools, and the broader community—with the goal of promoting the healthy acculturation, academic achievement, civic engagement, robust participation and leadership, and ultimately future success of refugee youth.

Is There a Magic Bullet?

Of course, there is no surefire single strategy, focus, program, or activity that guarantees the well-being, positive development, or success of refugee youth. According to a national panel of experts, however, the following features have proven to serve as reliable predictors of program effectiveness and as contributors to the success of youth:[32]

- Physical and psychological safety
- Appropriate structure

- Supportive relationships
- Opportunities to belong
- Positive social norms
- Support for efficacy and mattering
- Opportunities for skill building
- Integration of family, school and family, school and community efforts

While these features are crucial elements of quality youth programming, there are distinct layers requiring particular consideration when designing refugee youth/SIFE programs. As a general rule of thumb, IRC adopts a holistic and integrated approach involving all relevant actors to address the multiple issues impacting refugee youth, many of whom are SIFE.[33] With refugee youth at the center, targeted interventions radiate outward, drawing in the parents or guardians at the first level. Classroom teachers are next, followed by guidance counselors, school administrators, superintendents, and the board of education. Finally, IRC engages the broader community, including policy makers at various levels, to promote understanding and spur concerted action to tackle the challenges affecting refugee youth. To outline a few specific components:

Academic Support: Although there are overlapping SIFE and ELL needs, SIFE youth often require additional assistance to acquire fundamental skills such as phonetic/phonemic basics, decoding skills, and logical sequence in both the native language and English.[34] SIFE youth may also lack understanding of basic concepts, content knowledge, and critical-thinking skills.[35] The majority of IRC programs offer after-school, Saturday, and/or home-based tutoring services geared toward individualized instruction designed to impart fundamental skills, including basic literacy and numeracy, as well as assistance with homework assignments to ensure academic achievement.

Psychosocial: For refugee youth, cultural and language barriers are commonly exacerbated by the effects of war trauma. Horrific abuses witnessed, experienced, or committed by refugee youth can manifest as debilitating emotional blocks to the learning process. According to one source, it is estimated that from 11 percent to 75 percent of adolescent refugees have post-traumatic stress disorder (PTSD).[36] Meanwhile, studies have shown that one-third of refugee youth experience significant PTSD symptoms and anywhere from 4 percent to 47 percent of refugee adolescents suffer from depression.[37] In response, several IRC programs provide or facilitate psychosocial services for youth, as well as family members, delivered by licensed mental health professionals with previous experience serving refugees. Additionally, IRC offers a broad range of arts, sports, recreational, youth leadership, conflict resolution, and other activities aimed at promoting healing, collaborative learning, positive coping skills, and social and emotional development.

It is key to design evidence-based programming, linking research and theory to practice. While challenging to design/replicate tried and true refugee youth/SIFE programming due to scarce information, service delivery can be improved by way of a strong design, monitoring and evaluation (DM&E) framework and enhanced channels for information sharing. A logical framework that identifies goals, objectives, effects, outputs, major activities, and major inputs as a means of defining the overall strategy, as well as criteria for measuring performance outcomes through sets of indicators and means of verification, can serve as a crucial programming tool. Consistent data collection and analyses can produce more measurable results by which to evaluate the effectiveness and impact of a program or activities—which, in turn, creates a growing knowledge base on which approaches work and why.

The following list represents a number of **promising practices** espoused by the IRC network of youth programs:

Comprehensive & Integrated Programming

- Develop a positive youth development framework that advances the intellectual, physical, emotional, and social development of refugee youth in culturally and developmentally appropriate ways during adolescence as well as the transitional period to adulthood.
- Adopt an integrative approach that targets but is not exclusively limited to refugee youth; comprehensive programs serving both refugee and non-refugee youth can facilitate the positive acculturation of refugee youth.
- Design youth programs offering a wide range of activities—academic enrichment, computer instruction, arts, music, sports, drama, field trips, service-learning projects, and leadership development—in order to boost academic success as well as allow youth to discover strengths and talents in other areas, build self-confidence, and enhance social integration.
- Promote healthy habits (health and hygiene) and provide life skills training (e.g., home safety and financial literacy) to benefit the entire family.

- Conduct diversity trainings and implement programmatic activities that encourage social interaction among youth of different ethnic groups.
- For academic enrichment activities, divide groups on the basis of language and scholastic ability rather than age.

Engaging Refugee Families

- Bridge the cultural and linguistic gaps that exist between refugee families and local schools; help refugee parents understand the school system, requirements, academic standards and expectations; and create linkages and facilitate communications between refugee parents and school representatives.
- Engage refugee parents from the very outset by outlining expectations, roles, and responsibilities and by involving them in select programmatic activities.
- Reinforce the significance of parental involvement by conducting regular home visits to discuss the academic progress of as well as challenges facing their children.
- Design a family literacy component as part of comprehensive refugee youth programming.

Promoting Youth Participation & Leadership

- Solicit input from refugee youth and actively involve them in every stage of program development, implementation, monitoring, and evaluation.
- Encourage youth to assume substantive leadership roles in planning, managing, and evaluating program activities.
- Institutionalize youth participation in the program management process by creating a youth advisory structure.
- Recognize and respect the culture and religious practices of refugee youth and allow them to participate at their own pace.

Teaming up with Schools

- Build solid partnerships with schools by engaging principals and planning interventions based on input from educators; holding staff meetings that include teachers and project staff; and facilitating

written and verbal communications between schools, refugee families, and youth program staff.
- Collaborate with educators to develop solutions for refugee youth; assist in identifying appropriate classroom placements; provide assistance with mediation and crisis intervention; and facilitate coordination of parent-teacher conferences.
- Ensure that academic enrichment curricula and activities are coordinated and integrated with regular school programs.
- Request statistical data on refugee youth from schools and districts involved and advocate for evidence-based school interventions.

Developing Robust Partnerships

- A blend of both public and private partnerships is essential for the success of youth programs; work collaboratively with schools, refugee state coordinators, refugee service providers, community-based organizations, mutual assistance associations, religious groups, universities, volunteers, employers, health providers, mental health professionals, child welfare agencies, professional associations, and other key partners.

Recruiting Qualified Staff & Volunteers

- Recruit bicultural and/or bilingual staff and volunteers.
- Ground academic enrichment and other youth program activities in a nurturing, nonjudgmental, and safe environment.
- Involve caring and committed peers and adults as role models and mentors who are good communicators and can set appropriate limits while at the same time provide encouragement, recognition, and praise.
- Enhance program quality through regular training for staff and volunteers; provide ongoing professional development opportunities for youth program staff.

Recommendations

In keeping with this book's goal of stimulating a dialogue between the various collective actors concerned with the well-being of young refugees, we now offer them some suggestions that have emerged from our work experience.

To Key Federal and State Agencies

- The U.S. Department of Health & Human Services/Office of Refugee Resettlement should build on the success of Refugee School Impact Grants by increasing funding levels for these initiatives, as well as earmarking funds for comprehensive refugee youth programming and research projects to advance knowledge, policy, and practice in this field.
- The U.S. Department of Education should analyze the impact of the NCLB Act on refugee youth, as well as SIFE youth, at the state and district levels to help inform policy and preventive measures to ensure that refugee/SIFE youth are not left behind.
- Allocate federal and state funds for the following purposes:
  - Increasing refugee youth access to English Language Learning (ELL), including in-class and after-school activities linked to school, taking into consideration students' transportation needs.
  - Improving access of ELL students' parents to academic resources and school personnel by developing a roster of professionally trained interpreters for parent-teacher communications and by instituting accessible mechanisms for interpreter services and linguistic support.
  - Recruiting and training school personnel responsible for conducting linguistically and culturally appropriate (and nonverbal, as necessary) assessments of the cognitive and linguistic abilities and emotional and social development of ELL students.
- Key agencies at the national, state, and local levels should develop a multifaceted and systematic framework to respond to the diverse needs of refugee and immigrant youth and children. This should be integrated into agency planning processes and result in a concerted plan of action and broad-based implementation strategy with realistic and time-bound targets.

To Educators

- Design initiatives taking into account the broad intersections and challenges facing refugee, immigrant, SIFE, low-income, and minority populations in accessing educational resources with the objective of addressing the salient needs of refugee youth without marginalizing other affected groups.

- Establish collaborative networks and concrete communication mechanisms with refugee parents, refugee service providers, community-based agencies, and other local stakeholders to address the educational needs of refugee youth.
- Set positive developmental goals for refugee youth; formulate academic and psychosocial indicators and corresponding data collection and analysis systems; and maintain comprehensive statistics on refugee students, including demographics and student participation, achievement and behavior data.
- Institute measures for the systematic collection of disaggregated data on refugee/SIFE youth at the national, state, and local levels to determine the impact of the NCLB Act on refugee/SIFE youth and to help inform policy and programming at all levels.

To Refugee Youth Program Staff

- Actively engage refugee youth and incorporate their views in the design, implementation, monitoring, and evaluation of programmatic activities.
- Develop a positive youth development framework, integrated strategies, and effective M&E systems to advance the academic success, social and emotional development, and positive acculturation of refugee youth.
- Establish data collection and information systems and conduct research in order to identify vulnerable subgroups, design evidence-based programming at all levels, refine performance measures, and track progress toward specific goals.
- Address the gender dimension in all aspects of refugee youth programming, including data collection and analysis, especially when youth originate from cultures with different social norms and divergent gender roles.
- Cultivate strong partnerships with a broad range of stakeholders, and in particular local school districts and educators—assist educators to recognize and address the distinct needs of refugee/SIFE youth, coordinate youth program academic enrichment activities with regular school curricula, facilitate communications between educators and refugee parents, and inform educators of the broader service provider network supporting refugees.
- Develop collaborative networks to improve refugee youth and their families' access to integrated services, including coordinated referral and support services.

To Academicians, Researchers, and Practitioners

- Utilizing a variety of longitudinal, interdisciplinary, comparative, and other approaches, conduct rigorous research on the following topics:
  - Patterns of refugee youth academic engagement and performance over time taking into account gender, developmental factors, ethnicity, culture, language, religion, country of origin, refugee experience, and resettlement location.
  - Influence of educational background (formal and nonformal) overseas—trajectory from schooling in the home country through refugee flight and displacement—on learning for refugee youth resettling in the United States.
  - Impact of interrupted formal education on refugee youth learning processes, social and emotional development, and academic success in the United States.
  - Risk, protective, and resiliency factors for refugee youth behavioral and school adjustment and predictors of academic engagement and achievement.
  - Impact of high-stakes testing on newly and recently arrived refugee/SIFE youth.
  - Gender dimensions of refugee and resettlement experiences for diverse groups and implications for comprehensive and integrated youth programming stateside.
  - Effects of violence, persecution, displacement, protracted refugee situations, and migration on refugee youth and implications for policy and practice.

- Research, compile, and disseminate promising practices for these topics:
  - Educating recently arrived refugee/SIFE youth taking into account the effects of war and ensuing flight on refugee youth of different ages; gender; socioeconomic status; and ethnic, cultural, and religious backgrounds—as well as U.S. resettlement locations and circumstances.
  - Promoting the successful acculturation of refugee youth generally, as well as former child soldiers, youth subject to protracted refugee camp situations, SIFE, and other vulnerable refugee youth who resettle in the United States.
  - Advancing refugee youth participation and empowerment through academic support; vocational training; leadership initiatives; adult/peer-to-peer mentoring; arts, sports, and music activities; and community service projects.

- Responding to the range of psychosocial needs of refugee youth at various phases of intervention by stakeholders with different levels of expertise.
- Identifying Formulation Developing short-term and long-term interventions that address the wide-ranging needs of refugee youth populations in the United States.
- Strategies for identifying and responding to refugee youth demonstrating learning disabilities as well as high potential.
- Collaborative partnerships among refugee youth, parents, schools, and communities to promote cross-cultural understanding and advance the positive development of youth.

Conclusion

Refugee youth and their families have endured great hardship in their journey to the United States. They arrive in this country with great expectations, promise, and a profound appreciation for the freedoms and opportunities they have been afforded. Too often, however, their efforts to advance are frustrated by educational and other systems and standards that fail to recognize their unique needs, experience, and potential. The IRC calls for a renewed commitment on the part of the U.S. government to one of the most vulnerable and promising refugee populations—youth. It is our belief that targeted resources, interventions, and research will ensure that youth surmount challenges facing them and positively enrich their adopted country as other refugees have before them.

Notes

1. International Rescue Committee, Children & Youth Protection and Development Unit Programs for Children in Armed Conflict. Retrieved in October 2006 from http://www.theirc.org/what/irc_programs_for_children_in_armed_conflict.html.
2. Ibid.
3. Pub. L. No. 96–212, § 201(b), 94 Stat. 103.
4. A **"refugee"** is primarily defined in the Immigration and Nationality Act as any person who is outside any country of such person's nationality or, in the case of a person having no nationality, is outside any country in which such person last habitually resided, and who is unable or unwilling to return to, and is unable or unwilling to avail himself or herself of the protection of, that country because of persecution or a well-founded fear of persecution on account of race, religion, nationality, membership in a particular social group, or political opinion. INA § 101(a)(42)(a); 8 USC § 1101(a)(42)(a).

5. U.S. Department of State (2006). Some Facts About Fiscal Year 2006: 13, 115 children were under 16 years and 6,430 fell between 16 and 20 years. FY06 total refugee arrivals: 41, 279.
6. U.S. Department of State, U.S. Department of Homeland Security, and U.S. Department of Health and Human Services (2006). *Proposed Refugee Admissions for FiscalYear 2007: Report to the Congress.*
7. International Rescue Committee (September 2006). IRC National Resettlement Network of Youth Programs Survey Results.
8. International Rescue Committee (July 2005). Child and Youth Protection and Development Education Program Fact Sheet.
9. Suárez-Orozco, C., Suárez-Orozco, M. & Todorova, I. (2008). *Learning in a New Land: Immigrant Students in American Society.* (Cambridge, MA: Harvard University Press), p. 1.
10. Ibid.
11. Ibid., p. 28.
12. Interview with Christina Piranio, IRC San Diego youth programs manager, October 2006.
13. Interview with Meghan Brumby, IRC program manager, October 2006.
14. Interview with Christine Petrie, IRC deputy vice president for resettlement programs, October 2006.
15. U.S. Department of Education. Executive Summary of the No Child Left Behind Act of 2001. Retrieved in October 2006 from http://www.ed.gov/nclb/overview/intro/execsumm.pdf.
16. Interview with Ann Flagg, IRC Baltimore resource developer, November 2006.
17. Interview with Spencer Hall, IRC Atlanta youth and adult education manager, November 2006.
18. Suárez-Orozco, C, Suárez-Orozco, M. & Todorova, I. (2007). *Moving Stories: Academic Passages of Immigrant Youth* (Cambridge, MA: Harvard Press), p. 42.
19. McBrien, J. Lynn (2006). Educational Needs and Barriers for Refugee Students in the United States: A Review of the Literature. *Review of Educational Research,* 75 (3), p. 333.
20. Ibid.
21. International Save the Children Alliance (2006). *Rewrite the Future: Education for Children in Conflict-affected Countries* (London: Cambridge House), p. 1.
22. DeCapua, A., Smathers, W. & Tang, L. F. (2007). Schooling Interrupted *Educational Leadership,* 64 (6), p. 40.
23. Ibid.
24. Abraham, Salim. "Easing Foreign-Born Children into City School." (2006). Columbia News Service. Retrieved September 28 from http://jscms.jrn.columbia.edu/cns/2006-02-28/abraham-sifeprogram/Easing foreign-born children into city schools.
25. Interview with Nancy Doherty, IRC Boston youth program manager, October 2006.
26. McBrien, J. Lynn (2006). Educational Needs and Barriers for Refugee Students in the United States: A Review of the Literature. *Review of Educational Research,* 75 (3), p. 340.

27. Interview with James Lenton, IRC New York youth initiatives director, October 2006.
28. Suárez-Orozco, Suárez-Orozco & Todorova (2008). *Learning in a New Land: Immigrant Students in American Society* (Cambridge, MA: The Belknap Press of Harvard University Press), p. 30.
29. Interview with Dianne Browning, IRC Salt Lake City resource development specialist, October 2006.
30. Ibid., p. 345.
31. Interview with James Lenton, IRC New York youth initiatives director, October 2006.
32. Thaddeus Ferber, Elizabeth Gaines, and Christi Goodman (October 2005). Positive Youth Development: State Strategies. *Strengthening State Policy: Research and Policy Report.* National Conference of State Legislatures, p. 1.
33. For purposes of this chapter, in accordance with the New York State Department of Education, "SIFE" refers to students with interrupted formal education who come from homes in which a language other than English is spoken and who have entered a U.S. school after second grade. In addition, they have had at least two fewer years of schooling than their peers and function at least two years below grade level in reading and math.
34. DeCapua, Smathers & Tang (2007). Schooling Interrupted, p. 41.
35. Ibid.
36. Ellis, B. H. Trauma and Mental Health in Child and Adolescent Refugees. *Center for Behavioral Health.* Retrieved October 1, 2007 from http://www.nctsnet.org/nctsn_assets/pdfs/Child_Refugee_Mental_Health_Ellis8-07.pdf.
37. Ibid.

15

Reasons for Living and Hoping: Strangers No Longer

Julianne Duncan
United States Conference of Catholic Bishops,
Migration and Refugee Services

Introduction

In thinking about the meta-topic of this book, it will be useful to grapple with what we believe success is and how we know it when we see it. Migration and Refugee Services of the United States Conference of Catholic Bishops (USCCB/MRS) has been among the large service providers for immigrants to the United States for most of the twentieth century, both informally through churches and local efforts and formally through national program and advocacy on behalf of refugees and other migrants. On the basis of our experience, we have focused program and advocacy on services for immigrant children with and without families and from all parts of the globe. We have some ideas about what has been successful and what barriers are crucial to overcome. We also have some areas where we think our academic colleagues may be able to assist to refine the focus for services and advocacy on behalf of migrating children.

While there are many activities and strategies which we use throughout our program and advocacy efforts, I believe that we can distill our efforts to one key idea: a child or youth must have a sense of purpose, a reason for living and hoping. Without hope, nothing else works; with a sense of purpose, barriers can be overcome. Thus, the focus of USCCB/MRS is to advocate for, and to provide directly, services for refugee and other migrant[1] children so that they are no longer strangers among us but are connected to families and communities. As part of the family and community, their life and efforts matter to themselves and others—they are needed. They have hope for their future.

How Will We Know Success When We See It?

Children and youth who emerge into responsible adulthood are able to form families and contribute to their family and community. They are able to care for themselves and loved ones. They are able to *have* "loved ones."

At the most basic level, these are the factors we see when we recognize a "successful" immigrant youth-turned-adult. Thus, our efforts to help young immigrants focus on what we see as the ingredients to foster development of the healthy, integrated young adult. Others participating in the symposium that preceded this volume have highlighted different aspects of success, focusing on school success and to some extent on economic success. We see these other aspects of success as important, but in our work we emphasize underlying factors that may ultimately also affect educational and economic success.

How Do We Foster the Sense of Purpose, Reason for Living?

Children and youth migrate for myriad reasons. Some are fleeing with many refugees from large world-shaking issues of war and hunger. Others are migrating to alleviate individual and family issues such as the need to find work and send money to those left behind or, sadly, to look for family, someone to love them. Some enter the receiving society already imbued with a sense of purpose; others are unmoored, their worlds shattered.

As those who would provide a safe haven, how do we foster their sense of safety, of purpose, their reason for living? How do we receive them so that they are no longer strangers but have the help they need to enter their new society?

Background

The Catholic Church in the United States has been active in assisting migrants throughout its history. More than the Catholic Church in some other countries, the U.S. church has been a church of migrants, and its teaching and social service efforts have often been focused on easing the plight of the most vulnerable among migrants. When the Catholic bishops in the United States organized their first national Episcopal conference in 1920, one of the first departments formed was that for services to migrants: Migration and Refugee Services (MRS).

Since then, USCCB/MRS has been a leading agency in the provision of services to refugees and migrants in general as well as of specialized services to unaccompanied child migrants. As the largest refugee resettlement agency in the world, USCCB/MRS has resettled thousands of refugees and other migrants from all corners of the globe. Services are offered regardless of religious faith or other background to any migrant who seeks help; MRS is currently serving the largest number of Moslem and Hindu refugees resettling in the United States.

History

Since the period of disruption between WWI and WWII, children in various lands have been seeking a safe haven in the United States and finding assistance from the Catholic Church to do so. Children escaping alone from Germany and other areas subject to Nazi domination entered the United States without documents. Some of those children were aided by the local churches and the organized Catholic Church at the national level in the United States. As WWII drew to a close, the Catholic Church made a more systematic effort to care for children in place in Europe and to provide an orderly resettlement effort for orphans and other unaccompanied children. When Cuba was taken over by Castro, significant numbers of unaccompanied children left the country in a migrant child flow named "Pedro Pan." Thousands of those children found assistance in the United States through programs of MRS.

Vietnamese children who left their country in the infamous "baby lifts" that took place after the fall of Saigon were received and placed via MRS. In the aftermath of the communist takeover of Southeast Asia, thousands of children were alone and outside their home countries. The United States received and cared for those children with USCCB/MRS and Lutheran Immigration and Refugee Service (LIRS) as the two agencies that received and placed the children in specialized foster care programs. Haitian children who fled the repressive regimes in their country found (and continue to find) a safe haven through the efforts of MRS.

As war and social disruption affected other parts of the world, refugee children began to enter the United States in noticeable numbers, with large numbers entering from Africa. In recent years about 5,000 children and young adults from Sudan, the Sudanese "Lost Boys," arrived (see Gold, in this volume).[2] Again, MRS was in the forefront of identifying, processing, and placing the children and young adults. The placement and family reunification of refugee and other migrant children remain a significant activity of MRS.

As migration has developed and increased from Asia and Latin America in recent years, many populations of migrants have tried to immigrate to the United States but have not necessarily fit the categories that would offer legal status with the right to remain. Since the early 1990s, MRS has taken an active role in providing care and family reunification services for Chinese and other Asian children who need short-term care while their immigration cases are being settled. MRS and other offices of USCCB have taken the lead in efforts to assist Central Americans and other Latin Americans who have sought refuge from war and other disruptions in their countries of origin. Unaccompanied children entering the United States to join families, to seek safety, or to better their circumstances have been assisted by MRS through several means. Both on a national level and through local diocesan and religious efforts, migrating children remain a focus for policy, advocacy, and services.

Numbers

Since 1975, more than 800,000 refugees have reached a safe resettlement in the United States through USCCB/MRS. In general, the percentage of resettled children under the United States Refugee Program, arriving with and without families, ranges between 30 and 40 percent of arrivals. Thus, for refugees alone, MRS has assisted between 250,000 and 300,000 migrant children since the current formal resettlement program began in 1975. Other migrants in other flows have received services and help in other ways.

Some numbers may provide perspective. The USCCB/MRS system alone has served the following children during the last quarter of the twentieth century and into the twenty-first.

- Refugee numbers:

    All refugees: 800,000 since 1975
    Estimated refugee children: 30 percent, or 240,000, since 1975
    All refugee children since 1990: 97,000 (definite numbers)
    Unaccompanied children reunified with families: 6,000 since 1990
    Unaccompanied children for foster care: 7,000 since 1978

- Cuban and Haitian entrants:

    All Cuban and Haitian entrants: 65,476 since 1991
    All Cuban and Haitian children: 30 percent, or 2,000, since 1991
    Cuban and Haitian family reunifications: 1,114 since 1991

Unaccompanied children for foster care: 2,000 since 1990
Operation Pedro Pan: 14,200 unaccompanied children between 1960 and 1963

- Trafficked children:

    Child trafficking victims served: 1,000
    Unaccompanied child trafficking victims cared for in foster care: 35 since 2001

- Undocumented children:

    Best Interest Recommendations/reunified with families: 13,000 since 2003
    Undocumented children reunited with family and provided specialized services: 1,002 since 1997
    Undocumented children cared for in foster care: 106 since 1997

Advocacy and Program Services

USCCB/MRS advocates on a national level for better government response to migrant children. National program activities provide specific responses to the needs of migrant children both within their families and with special attention to the safety net for children without parents.

Within refugee resettlement and programs for Cuban and Haitian entrants, resettlement sites provide services to stabilize families in welcoming communities. For children who are traveling with nonparental relatives or children reuniting with parents, specialized suitability assessments and follow-up services are arranged. When family breakdowns occur, specialized foster care services are available to pick up the pieces.

For migrant children traveling alone and for whom family members cannot be found, specialized foster care programs receive refugee, entrant, trafficked, and undocumented children. MRS-affiliated programs exist in ten states and receive children from any part of the world.

Through its Safe Passages program, MRS ensures that children traveling without adult relatives and without legal travel documents are provided a safety, and receive appropriate care while in the custody of the U.S. government. For children smuggled or trafficked in and at risk from organized crime, MRS works with local diocesan offices to determine whether family reunification is a suitable option, and follows up with the children to ensure their well-being.

The Bridging Refugee Youth and Children's Services (BRYCS) program supports service providers for refugee youth, children, and families through targeted local trainings, development of original resources, and access to information resources through its Web site and online clearinghouse. Since its launch in 2001, BRYCS has provided more than 170,000 downloads of documents that assist the child welfare system, resettlement system, or refugees themselves. Onsite or other in-person technical assistance is also available on www.brycs.org. It is our intention to have that clearinghouse serve as institutional memory such that resettlement efforts need not be reinvented with every new group of refugees and immigrants who enter the United States or other countries.

Victims of human trafficking are forced into inhuman, slave-like conditions as workers in the sex trade, sweatshops, and domestic labor industries. This contemporary slave trade has become as lucrative as the arms trade. Each year, up to two million people worldwide are victims of trafficking, including 20,000 men, women, and children who are brought into the United States, one-third of whom are under the age of 18. Traffickers often lure victims into the country under false pretexts and take away their documents to trap them.

Since 2002, MRS has led efforts to combat this modern-day evil through advocacy, increasing public awareness, training and technical assistance to service providers, and direct outreach to the trafficking victims themselves. MRS places trafficked children into foster care, group homes, or independent living arrangements and monitors their care and well-being. USCCB/MRS has participated with Georgetown University, Institute for the Study of International Migration, in a study of child trafficking victims who have received benefits in our own and other programs. Results of that study inform us about the needs of the victims as well as which services are more effective with the specific population.[3]

Has There Been Success; How Do We Know?

We think the effort to receive and integrate immigrants has been a huge success in the United States over many years and that the refugee and other migrant children and youth who have come to our shores have indeed become responsible contributing adult members of society. Some of our academic colleagues at the prepublication symposium identify the existence of "high quality resettlement services" as a factor in the success of the populations they have studied and discuss in this volume. For example, the research on the Sudanese children and youth studied by our colleagues from Michigan State University and included in this volume was conducted on populations resettled by USCCB/MRS and the Catholic social service system and is one of few studies we are aware of that documents the beneficial effect of resettlement social services.

Some research identifies "best practice" and supports efforts to focus program in one direction or another. Within social services, the Council on Accreditation analyzes what is known about best practice, and agencies receiving accreditation deliver their services in accordance with COA standards for Immigration and Refugee Resettlement.[4] Within the USCCB/MRS referral system, most Refugee Resettlement and all services for Unaccompanied Refugee Minors are delivered by accredited agencies.

In other situations, research is less clear about effective program design. Nevertheless, we design programs on the basis of our best information about what helps in migrant integration and continue to adjust efforts as conditions change. With a system as large and that must adjust to changing conditions, we do not have the quantitative information that we would like to have in order to continually refine our practice.

What Helps?

As we design and carry out programming for migrant children, we have identified a few key factors that seem to assist in child and family integration.

Supportive Structures Must Be in Place for Children to Thrive and Succeed

Family
Migrant children and youth find it easier to succeed if they have a supportive social group. If they are able to migrate with an intact family, they are able to grow up in a familiar structure in which they are known and needed. For those children who migrate without their own family, maintaining whatever connection possible with family is immensely helpful. When children are truly alone or their families are unable to care for them, creating an alternate family provides essential connection tethering the child to the universe. Therefore, USCCB/MRS provides specialized services for unaccompanied refugee, asylee, and trafficked children in foster care programs in ten locations in the country. Unaccompanied children are placed directly in the foster care programs. Families in which the caregiver is elderly or ill or for other reasons may be unable to maintain care for the children in the family unit are placed in locations near the foster care programs so that if the children need that service on either a temporary or permanent basis, the child and family may maintain bonds and will experience the least disruption possible under difficult circumstances.

> Two small children, six and eight years old, entered the United States with their elderly grandmother who had been caring for them in an African refugee camp since their mother's death. On the basis of the age and poor health of the grandmother, we determined that this family unit was fragile and that there must be a plan for the children's care should she be unable to maintain a household. The family was placed in a resettlement program in the mid-west near a refugee foster care program. Staff from both programs cooperated in services to the grandmother and the children. Upon her death approximately a year after arrival, the children entered the foster care program to live with a family who had begun to get to know them by providing tutoring while they still lived with their grandmother. They remain in care with their foster family and have established correspondence with distant relatives in their country of origin. They do some of their school projects on the basis of their country of origin and will have an opportunity to visit if conditions of peace develop in the future.[5]

Community
In addition to family, a sense of community provides the buffer between the smaller unit and the larger society. Where there is a community of people from a similar background, children and their families have more sources of strength and support to assist in adjusting to the vagaries of their new society.

Therefore, the agency maintains the refugee foster care programs in communities in which there is a strong resettlement program and tries to place unaccompanied children in communities in which adults and families of the same background are already living or are being resettled.

Where there are few other people of familiar background in potential resettlement communities, the structure of the resettlement agencies, churches, or other assistance systems can create communities and safety nets that will allow the natural resilience of the child, youth, family, and community to take root and thrive.

Strengths Find Outlet in New Location

Where the receiving society obviously needs the strengths, talents, or skills that the migrant child or youth brings, integration is easier. The youths can find a niche more easily and can know themselves as contributing members of the new society. Somewhat less learning and adjustment are needed. Where the strengths and gifts the youths bring are less obvious, the receiving agencies, churches, and other welcoming groups must play a larger role in assisting the youths to identify the contributions needed by their new community.

Faith; Welcoming Church

For those children and youth who have a sense of faith or religious belief, that belief may provide their sense of purpose and may make it possible to find a supportive, welcoming community in their new home. Local churches and other religious groups may need assistance to find the migrants among them and to understand how to reach out in welcome. As many refugees and other migrants now entering the United States are of different faiths than many in their receiving communities, the resettlement system must and does work with all faith communities to receive and resettle refugees.

> A sibling group of five children and one young adult entered the United States from a refugee camp in South Asia. The USCCB/MRS providers in the southern state where they were placed arranged for the group to live with a Moslem family from India where the young adult could begin his adult life with job training and the assistance he needed to get on his feet. The children entered school and community activities. In the community were two mosques; the young man and the older children began participating in religious services and social events. Eventually, the young man left the foster home and lived in a small house owned and managed by one of the mosques and, though he struggled, began to achieve self sufficiency. Several years later, the oldest of the children are graduating from community college and the young men of the family are creating a household. The younger children and girls are still residing in foster care.

Future Orientation

Since immediate comfort and financial security after migration are rare, those children and youth who have a time sense that focuses on the future and allows for measured steps toward a distant goal are more easily able to delay gratification and persist in efforts to learn and adjust. For those who are present focused, agencies and programs may effectively work to reframe, redirect them to longer-range goals with guidance on intermediate steps. Since so many refugees now entering the United States have lived many years in refugee camps where they needed a "present" orientation to survive, specific intervention is sometimes needed to assist refugee youth to think beyond their immediate needs. The web-based clearinghouse and other technical assistance services, provided by USCCB/MRS through the BRYCS program, offer resources for mainstream and refugee agencies to use to identify promising practice with refugee and immigrant youth. Recent publications of a "Positive Youth Development Toolkit" and a "Cross-cultural Parenting Toolkit" appear to be useful to providers throughout the country.[6]

What Inhibits Success?

Anomie; No Social Bonds

Children and youth who remain unmoored from a healthy family and community more easily fall into gangs and criminal groups, which buffer them from the anomie and stress of social adjustment. Children without bonds to kindly adults or safe community structures will find a sense of meaning and belonging wherever they can. The children who fall prey to traffickers are often enticed into their situations and remain with their traffickers because they believe the trafficker loves them.

Family Breakdown

When the family buffer shatters, children are left stranded. Where children have migrated with, or settled with, nonparental relatives, bonds may be especially fragile, and children have few safety nets if the family system does not remain intact. Even if the family remains intact but the adults are out of the home working, children and youth may become isolated from their family support system.

> A sibling group of three young adults and two mid-teen children were resettled together in Florida about five years ago. The young adults began working within a short time and found jobs mainly in grocery stores where they were needed to stock shelves and do work that placed them on the job in the evening hours when the teens were

> home from school. The family unit eventually fractured with the teens beginning to get involved with antisocial activities. The resettlement office intervened to assist the older family members to structure their work and household activities in such a way that they developed a cooperative household rather than one in which each member was too busy to interact. As they have remained in the United States longer, they have adjusted their routines and interactions to accommodate each other's activities. The youngest of the sibling group has just graduated from high school.

Age At Time of Migration—Various Implications

The age of a child or youth at migration affects their adjustment in ways that make their ultimate success more or less difficult. Those who migrate at younger ages may absorb the language and culture of the receiving society easily, thus giving them the benefit of an easier adjustment to the wider world. On the other hand, those who have reached mid-adolescence before migrating may have been among the fortunate ones who had a stable upbringing in their society of origin and who enter the new society with well-integrated values.

A troubling, more recent phenomena for refugee children is the grave problem of children who remain for many years in refugee camps without being able to reunite with their families but without being able to access any other durable solution either. If such children do eventually migrate to the United States or other receiving countries, they have often missed crucial developmental milestones and struggle to integrate emotionally and socially.

The resettlement of a group of children and now young adults from southern Sudan during 2000–2002 illustrates the situation of many children in protracted refugee situations. As children, they left their homes in the late 1980s, fleeing first to Ethiopia and eventually to Kenya, where they entered United Nations High Commissioner for Refugees (UNHCR) camps in 1992. Many were as young as four or five years old. Those children and then young adults remained in the camps until their resettlement, having stayed in refugee camps with harsh conditions for up to ten years of their young lives. Yet, the cohort appears to be resilient, with relatively few displaying severe symptoms of distress. For example, one year after arrival, only 20 percent scored in the diagnostic range for PTSD, despite virtually all having witnessed or been victimized by war-related violence.[7]

Nevertheless, because of the wait in the refugee camp they were not able to achieve permanency throughout most of their developmental years and bear emotional scars from that long stay. Follow-up studies of the research cohort may reveal more nuanced information.

Lack of Legal Status a Barrier

One aspect of recent migration to the United States as well as to other areas of the developed world appears to us to be adversely affecting young migrants. Anti-immigrant sentiment and corresponding laws make it difficult to achieve legal status and thus are forcing children and youth into an underground where they are unprotected and more easily preyed upon by the unscrupulous. With barriers to legalization growing higher and higher, we are developing an underclass of people who cannot integrate into the receiving society and are prevented from becoming contributing members of civil society (see Bean et al., this volume).

Broad Social Attitudes That Affect Migrant Children and Youth

During the last several years we have seen a less generous, less welcoming attitude on the part of the United States and other countries, which affects children significantly. Children without families to advocate for them are least able to navigate the barriers to care. They more and more easily fall through the cracks in the refugee system or in other immigration systems and are denied entry. If they have gained entry, they may again slip through the cracks of the domestic refugee and child welfare systems.

In the United States currently, a combination of anti-immigrant and antigovernment sentiment has inhibited integration services for children and youth. We see services being offered in a more punitive manner, with greater attempts to exclude children from care than in the recent past. Fewer children are granted asylum than in the past. Children in federal custody because of immigration status are cared for in a care model that uses less of the successful foster care programs and more congregate care emphasizing "control" despite efforts at improvement. Children who would appear to be eligible for foster care are excluded in ways that would not have been likely in the past.

For refugee children not yet in the United States but for whom resettlement should be considered, we see that barriers to inclusion are greater and greater. The issue referred to as "material support" has affected all resettlement, including that of children. Anti-resettlement attitudes in much of the world have prevented children from participating in some resettlement efforts even when others from the country are able to move. The most vulnerable unaccompanied children may be left behind.

The inability of many current migrants, including children, to achieve legal status in the United States has created and intensified a two-tier society, with an underground society and economy that has chilling effects on children and youth. If they are unable to find a haven in their country of origin, they are now also often not able to find it in the United States. Therefore, the youth cannot enter adulthood as contributing members of their new society.

Program Design Implications and Recommendations

To encourage those factors that seem to be helpful and to diminish those that hinder, we have learned from experience that program design should adopt several principles.

Family Cohesion When Known and Possible

Whenever possible, maintain family unity so that children remain with or are reunited with parents or other trusted adults. Balancing family unity and child safety from possible abuse or eventual abandonment after resettlement remains a constant and delicate challenge. When children are resettled alone, programs must undertake to create substitute family and community bonds as soon as possible.

Resettle in Clusters to Create Community

When resettlement is deliberate, as with refugees and entrants who are assigned to locations and assisted upon arrival, creating communities by clustering people in specified locations assists in the building of long-term support systems.

When resettlement is spontaneous or irregular as is the case with many undocumented migrants and other immigrants, the voluntary agency sector may assist in the creation of organizations that can enhance a sense of community. Any means of creating community—churches, mutual assistance associations, business organizations—will help the children and youth.

MRS resettles refugee and entrant families in carefully identified locations through diocesan programs that emphasize community and parish support. Ethnic mutual assistance associations are frequently housed and assisted by the local Catholic Charities until they are able to thrive on their own. Organizations and programs to assist parents and children are the initial focus of much local programming for newly arrived refugees.

Special Assistance for Children Separated from Parents

Children who resettle with nonparental relatives or travel to join parents or other relatives are at greater risk than children traveling in intact families. Special suitability assessments and follow-up services assist refugees, entrants, and some undocumented children at greatest risk for harm. MRS has specially designated national staff to assist local sites when "attached[8]" minors are placed or children are joining parents after periods of separation.

MRS recognizes the need to create safe havens where children can turn for help when their safety net frays. Of particular concern are those children who exist in the underground economy and who are therefore at the greatest risk of being trafficked. Designated national staff are available to local sites who find that a resettled refugee child or other fragile children are in need of entry into the unaccompanied refugee minor foster care programs because their family system has not been able to care for them.

Create Families and Communities for Unaccompanied Minors

Children traveling without families and who have no families available to care for them need the assistance of specialized programs to create families and foster their connection to their ethnic community as well as the larger society they are entering. The successful efforts of the Unaccompanied Refugee Minors programs have assisted thousands of unaccompanied refugee and other migrant children to integrate into families and communities throughout the United States since 1978 when the programs were first formed.

Community Integration—to Both Ethnic Communities and Geographic Communities

All migrants need help to find a welcoming community through agencies, ethnic communities, churches, or religious groups and whatever aspects of civil society can become available to receive them. MRS works through more than 100 local Catholic dioceses to make initial placements and to assist the migrant to find or form ethnic associations. Often ethnic self-help organizations are initially housed and assisted by the diocesan resettlement program.

What We Do Not Know

We are aware of the successful lives of many of the refugee children who have participated in our programs, some of whom come forward to become foster parents or sponsors for later arriving refugees. However, we do not have scientific studies of long-term outcomes. The question remains open, especially for those children without families: are the deliberate efforts to assist refugee and entrant children leading to more successful outcomes than the ad hoc arrangements that undocumented, unaccompanied children and youth find for themselves? Efforts from our academic colleagues to examine this question for unaccompanied children entering the United States or other destination countries would be useful (see Sonnert and Holton, in this volume). Examining differences in practice and outcomes among resettlement countries may also be of use to those of us trying to formulate good program assistance.

What is the difference in success between the refugee children who migrated or resettled versus the ones who did not? Studies of those children and youth who migrated compared with a comparable group who did not would be of interest. Considering the large numbers of unaccompanied refugee children and youth who are not offered an opportunity to leave the refugee camps, it should be possible to design a study that would compare those who had an opportunity to resettle with those of the same or similar background who either remained in refugee camps or returned eventually to their country of origin after long stays in the country of first asylum.

Conclusion

We appreciate the opportunity to participate in the efforts that have led to this volume, and would be interested in cooperating in research efforts of scholars who are looking at the questions raised here. Future studies and possible collaboration will provide more information and should prove useful in program design.

In the meantime, we hope that the United States may renew its welcome to refugees and immigrants and take pride in the individual success of those children and youth who have entered American society and contributed to the success of our country.

Notes

1. The term "migrant" is used to denote those children who enter the United States who may or may not have legal immigration status. In keeping with common and legal usage, "refugee" and "immigrant" have specific definitions. "Migrant" encompasses any person moving from one country to another. Usage differs somewhat between the United States and European and other countries and has changed over time as U.S. laws have been enacted and international conventions have been signed. Refer to www.brycs.org for definitions of terms.
2. The young adults and some of the children referred to in this volume who were resettled in Lansing, MI, were resettled and served by USCCB/MRS and its local partner.
3. Gozdziak, Elzbieta M. and Micah Bump (2007). *Victims No Longer: Research on Child Survivors of Trafficking for Sexual and Labor Exploitation.* National Institute of Justice, Washington, D.C.
4. See www.coastandards.org for the Standards of Excellence for Immigration and Refugee Resettlement and the supporting research upon which the standards have been based.
5. All vignettes in this chapter are actual cases of resettled refugee or immigrant families. Locations and minor details have been changed to protect their privacy.
6. These and other resources are available at www.brycs.org.
7. Geltman, Paul L., Grant Knight, Wanda, Lloyd-Travaglini, Christine, Lustig, Stuart, Landgraf, Jeanne M., Wise, Paul. *A Nationwide Study of the Functional and Behavioral Health of Sudanese Unaccompanied Refugee Minors Resettled in the United States. Executive Summary.* April 2003. Available on www.brycs.org.
8. "Attached minors" is the term used in the United States Refugee Program to refer to children traveling with and resettling apart from their parents or traveling to join parents. In other refugee contexts, the term frequently used is "separated child."

16

Protection of Refugee and Migrant Children: The Role of Nonprofits

Annie Wilson
Lutheran Immigration and Refugee Service

Massive worldwide migration is a defining hallmark of our current era and is likely to swell even more in coming years, as the forces that drive and support migration are strengthened. These forces include economic globalization and the existence of global labor markets, access to communication networks and the increasing density of social networks across international borders, the availability of transportation, political instability, and environmental factors.

From the societal perspective, migration carries with it both benefits and challenges, making it a political hot button issue. No nation is untouched. Across the globe, societies are struggling to successfully absorb newcomers, creating or reforming legal frameworks to address both the regulation of migration (enforcement) and to protect the human rights of migrants, attempting to maximize the economic benefits of migration, dealing with ethnic or cultural conflict arising from the presence of migrants, promoting integration, and meeting the needs of their own nationals living in other countries.

Within the broad phenomenon of migration, there are important distinctions. The term "migrant" can be used in an encompassing sense to refer to anyone who is on the move—whether voluntarily or involuntarily, and regardless of reason: economic, political, or personal. It is in this broader sense, rather than in the narrower sense sometimes employed of an individual seeking work in a new land, that the term is used in this chapter. An "immigrant" is someone who has voluntarily chosen to migrate, and may be documented or undocumented (legal or illegal). A "refugee" is someone who has been forced to leave their country of origin because of a well-founded fear of persecution. In the United States, people who have official refugee status have been brought here under a special program with partial government funding. Some refugees, however, arrive in the United States either without documents or under some other immigration status, and request asylum. While the request is pending, these individuals are "asylum seekers," and if the request is granted, they become "asylees." Refugees and asylees are eligible to become U.S. citizens after a period of time.

Regardless of the particular migration status they hold, children deserve additional attention and require special protections because of their age. Many children are members of migrating families, or are significantly affected by the migration of one or both parents. In the United States, 13 percent of all children (9.1 million children) are members of families that include both citizens and noncitizens. Children also comprise a high percentage of the worldwide refugee population.

Children who migrate alone, because they have become separated from their families or are without family, are particularly vulnerable. Nine thousand such children are held in the custody of the U.S. federal government each year.

Whatever their situation, migrant and refugee children face challenges and have vulnerabilities that are distinct from those faced by adults.

The Role of Nonprofit Organizations

To successfully address highly complex social issues such as migration, the coordinated engagement of multiple sectors of society is required, as no one sector holds all the answers or can single-handedly implement solutions. Among the sectors that are critical for addressing migration concerns are government, the nonprofit sector, the academic sector, and for-profit corporations. This chapter will highlight the complementary yet distinct role of nonprofit organizations in the protection of refugee and migrant children.

Nonprofits are perhaps best known for their work in carrying out the delivery of direct services. Nonprofits have many unique strengths in the area of service delivery, including a deserved reputation for delivering service with a "heart," a holistic approach to clients, expertise built up across institutional structures over many years, close connections to local communities, and the ability to engage and motivate volunteers. But service delivery is only part of the story. Because of their distinctive organizational structure (in particular, governance by boards of directors representing concerned communities), and their institutional accountabilities to constituencies, nonprofits are driven by a sense of mission. They are focused on the long-term future; they try to bring about change; and they act as a voice for those they serve. The government sector, while certainly sharing these concerns, is driven by its own structure and accountabilities to a focus on creating and adhering to an overall regulatory structure, promoting principles of fair access, and near-term planning.[1]

With regard to the protection of refugee and migrant children, nonprofits play multiple roles. In the delivery of direct services, both here in the United States and around the world, they care for children in refugee camps, provide legal services and foster care, assess the best interests of children and the capacities of caregivers, and provide a wide range of social services that help with the integration process. For the sake of efficiency in oversight or more centralized control, such services could certainly be delivered directly by governments or even by for-profit entities, although arguably with the loss of the many positive qualities that nonprofits bring to the service delivery role (notably, a commitment to individualized care). However, nonprofits play an additional role, beyond direct service delivery, that is not replaceable. They ensure that the overall framework of protection for children is continually strengthened through their engagement and leadership in the process known as "societal learning,"[2] in which multiple sectors of society, working together in partnership, gain knowledge of the issues, the best strategies for addressing them, and the most effective ways of working together.

In the United States, budgetary pressures and recently shifting philosophies of service delivery within the administration point to the possibility of a diminishing role for nonprofits in federally supported work with refugee and migrant children. The focus of the tension has been the nonprofit role in direct service delivery. However, sidelining nonprofits without taking into account their broader role in bringing about needed systemic changes and building the capacity of communities to meet societal needs will have unintended consequences, weakening our collective ability to sustain and improve the framework of protection for migrant children.

Background on Lutheran Immigration and Refugee Service

The Lutheran Immigration and Refugee Service (LIRS) was established in 1939 to meet needs of refugees from Germany, and the organization's core mission of welcoming the stranger through ministries of service and justice has informed our work since that time. Over the decades, we have been involved in a range of services to displaced people, immigrants and refugees from around the world who have hoped to build new lives in the United States. Starting in 1975, LIRS has participated in a public/private partnership with the federal government, and has received federal funding to carry out many activities, particularly in the area of refugee services.

In addition to refugee resettlement and continuing services to refugees, which LIRS carries out primarily through its network of affiliates, our organization provides financial and technical support to about 25 community-based legal programs serving asylum seekers and immigrants in detention; supports and coordinates a network of legal services to survivors of torture in detention; and provides significant technical assistance in the areas of refugee employment, immigration law, and institutional capacity building. Our organization believes that service and advocacy should be inseparable, and has committed increasing resources to communications, public education, policy and advocacy.

LIRS traces its special commitment to children to the years immediately following World War II, when an LIRS social worker named Cordelia Cox, working in camps in Europe, vigorously promoted the interests of women and children. Along with our close colleagues at the U.S. Conference of Catholic Bishops, LIRS is one of two agencies that places unaccompanied refugee children in foster care, and we serve separated and unaccompanied migrant children in the custody of the federal government, in partnership with the Department of Health and Human Services' Office of Refugee Resettlement. Through our programs, and as advocates in the national and international arenas, LIRS promotes policies and practices that are centered on the best interest of the child.

A Framework for Success

For the past couple of years, the nine national agencies involved in refugee resettlement have been engaged in developing standards of success for our work. On our own initiative, we have been considering not only what constitutes successful resettlement in the short and long term for refugees themselves, but also what constitutes success for host communities, for governmental and nongovernmental institutional structures engaged in services, and for the humanitarian and legal framework within which we work. We are beginning to test our thinking with data collection instruments we have developed, and as we refine our work we hope to engage governmental partners and other stakeholders in some new thinking about the programs we carry out together.

Our work describes what success looks like for nonprofits engaged in work with newcomers, and helps with an understanding of the unique role of nonprofits relative to other players. The standards of success were not specifically designed for children, although the well-being of children and the ability of adult caregivers to promote child welfare within the U.S. context were identified as discrete elements of success within our broader framework. The broad concepts of the standards, while designed for the specific context of the refugee resettlement program, are nevertheless applicable to work with refugee and migrant children.

In the sections that follow, the overall framework of success that the national resettlement agencies are developing is presented, with some discussion of their applicability to refugee and migrant children, highlighting the ways in which children's needs differ from adults'. This chapter touches on some of the ways nonprofits respond programmatically, and suggests challenges and recommendations regarding the role of nonprofits in protecting children.

At the broadest conceptual level, the resettlement agencies identified three domains for success. These are protection, stability, and integration. Protection encompasses the existence of a legal and humanitarian framework responsive to the needs of refugees, strong public support for humanitarian work and acceptance of newcomers, meaningful access to legal protection, and practical physical safety and security. Stability refers to the short-term process of overcoming the life disruption attendant on arrival in a new country as a refugee. A refugee who has successfully achieved stability has his or her basic physical needs met, has procured legal protection, knows where to turn for assistance with life emergencies, and is ready to begin the long process of integration into a new society. Integration is the long-term process in which successful refugees become increasingly active participants in the economic, social, and political spheres of their new land, learning how to contribute their unique talents and perspectives and finding the appropriate balance between old and new worlds.

Migrating Children—A Variety of Circumstances

Children migrate under a wide range of circumstances. Important distinctions include whether children migrate with legal status or without, and whether children migrate with parents or customary caregivers, or alone. Particularly vulnerable categories of children include those who are refugees (with or without recognition—although those who are not yet legally recognized as refugees [i.e., asylum seekers] may face additional peril), children who are separated or unaccompanied, children who are trafficked, children who are without home or caregivers in the country of origin, children who are experiencing domestic abuse or neglect, and children who have experienced trauma at any point during the migration experience.

Although the particular interventions that are needed will vary for each of these groups, success for all migrating children requires that they be protected (as both migrants and children), that at critical life junctures they stabilize their immediate situations, and that they be able to proceed along a path to integration in their country of residence. The specific milestones for success for children need to accommodate their developmental requirements, recognizing their participation in family life, school, and community and reflecting their orientation toward the future.

Protection

Protection is the first of the domains identified in the standards of success, and is the domain in which the role of nonprofits is least understood. It should be noted here that "protection" is a term of art in the refugee world, with specific legal meaning and practical implications. In this chapter, as in the work being carried out by the national resettlement agencies, the term is being used in a more general sense, although with reference to the thinking that has gone into the refugee terminology. Three important forms of protection for children are physical care and protection, legal protection for the individual child and adequate legal and institutional systems.

All migrating children, regardless of circumstances, require physical care and protection. A child who is physically protected is safe from danger and violence, is free from privation, and is not exploited for labor, sex, or any other purpose. The child himself or herself understands (at an age-appropriate level) the right to be physically protected and how to access remedies. For most children, physical protection is provided by parents or customary caregivers, and when the physical protection of refugee or migrant parents has been secured, they in turn are able to protect their children. It is, however, important to note that even children accompanied by parents or caregivers may face child-specific risks. It is critical that these be considered and addressed by states or organizations assisting migrants and refugees, and that it not be assumed that accompanied children have no special protection needs.

The successful physical protection of unaccompanied or separated children, as distinct from those accompanied by their caregivers, is a serious challenge, and requires that some entity fill the role of the parent or customary caregiver until children can be reunited with their families or alternative care arrangements established. In the chaos of flight or migration, arrangements for such children may be absent or ad hoc. Some children are under the physical care of adults who abuse or exploit them. The successful protection of separated or unaccompanied children requires that an individual determination of the child's best interests be carried out at the soonest possible juncture, and that there be appropriate care-giving options available, with follow up services provided. Currently, a lack of resources prohibits "best interest determinations" and other services from being carried out timely for many separated and unaccompanied refugee children residing in camps. Full implementation of a program of comprehensive services for separated migrant children in the custody of the U.S. federal government has also not been possible to date, due to resource and infrastructure challenges, although progress is being made.

Legal protection for refugee and migrant children is a critical component of their success. Without legal protection, children's lives are in limbo, and planning for permanency is impossible. Many migrant children in the United States are caught up in a system whose incompatible goals deny them a long-term legal solution. Immigration enforcement goals undercut access to the legal protections that would otherwise be available, creating both substantive and procedural barriers. Even where legal protection has been extended for the short term (as is the case for recognized refugees living in camps), the absence of a viable permanent legal status resulting from the implementation of durable solutions is extremely detrimental to proper child development. The shameful international practice of refugee warehousing denies children a future.

The successful protection of refugee and migrant children requires that each child receive physical care and protection, as well as legal protection in both the short and long term. In order to ensure that these are in place for the individual child, it is necessary that there be a framework of healthy legal and humanitarian systems, supported by the state and the public at large. Such systems can only be created, sustained, and advanced with the involvement of multiple sectors of society, including the academic world, civil society, and the public sector. The engagement of the academic world ensures that there is rigor in the system, and that legal and humanitarian practices directed toward refugee and migrant children are actually resulting in the best outcomes. The engagement of civil society (non profit organizations and the philanthropic community) brings the power of public advocacy to the issues. Committed constituents draw from firsthand experience in service to be a voice for change, and their perspective is personal, passionate, holistic, and future oriented. The public sector (government) brings to bear financial resources no other sector can command, establishes the regulatory structure within which work is carried out, and promotes a culture favoring fair and equal treatment.

In the case of refugee and migrant children, it is not only different sectors of society that must find ways to work together effectively, but different systems of law and humanitarian support must interact. To give one example, refugee children and their families resettled in the United States are sometimes caught between the child welfare system and the refugee-serving system. The child welfare system in many places lacks familiarity with the culture and background of refugees and is unaware of good practices in serving non-English speakers. Similarly, the refugee-serving system may be unaware of how the child welfare system works, so that refugee families inadvertently violate U.S. child welfare practices. The data are suggestive that comparatively high numbers of refugee children are removed from their families by child welfare authorities, where greater coordination and information sharing may have kept more families intact.

In a second example, lawyers and child welfare workers occasionally clash over the appropriate course of action for particular children. In their role as zealous advocates, lawyers may pursue legal avenues for a child to remain in the United States, while child welfare workers advise return to the country of origin and reunification with family. Nonprofits and government, working together on programs serving children, also experience difficult communication at times. Nonprofits' advocacy and desire for systemic change is experienced as disruptive by federal administrators, and it is sometimes perceived that nonprofits are acting in self-interest when they advocate for more services for clients, making it difficult to address the underlying issues at stake. In all these cases, such differences in professional opinion, arising from the legitimate perspectives of different systems, can create animosity and mutual suspicion at the interpersonal level.

For protection to be available to individual refugee and migrant children, there must be a continuous, deliberate, and dynamic process of interaction across sectors and systems. Nonprofits and the philanthropic community play a critical role in ensuring that such interaction takes place, and that the public at large is informed and engaged, providing the necessary political support for programs of service. Avoiding interaction across sectors and the difficult conversations sometimes entailed—whether it be between lawyers and social workers, refugee-serving systems and child welfare systems, or government and nonprofit organizations—ultimately fails children.

Stability

Quite often, nonprofit organizations like LIRS are engaged in the provision of relatively short-term services that serve to provide a measure of stability to refugees or migrants at a critical life juncture. Examples of such services include the reception and placement services of refugee resettlement, the placement of an unaccompanied refugee minor into a foster care program, or legal services to a migrant facing removal (deportation). Stability is a precondition for the longer-term process of integration, but the markers of success within this domain are distinct. It is unrealistic to expect that the achievements of integration will take place during the resolution of a major life event.

What are the markers of success in stability? An individual who has achieved stability does not fear a sudden change of legal status that will turn life upside down. He or she can count on food and shelter being available into the near-term future, and is capable of basic self-care, having mastered, for example, the basic tasks of life in a new community (acquiring and preparing food, securing transportation to critical locations, a vocabulary of survival English). A stable individual has a reasonable plan for achieving economic, social and political integration at some point in the future, and some optimism looking forward.

Most of the direct human services we provide to refugee and migrant children through our national office at LIRS are intended to help them achieve short-term stabilization, rather than long-term integration. We assist with the process of family reunification for children and their parents or other caregivers, and we place children in foster care programs. We serve children as members of refugee families, providing them with three months of reception and placement services. We provide legal counseling and representation to children involved in immigration proceedings. We link children to the service providers who will be assisting with the process of integration. Frequently, our services are not enough to ensure stability for children who are undocumented migrants, because their legal situations remain unresolved. Successful long-term integration for migrant children, or any form of permanency planning, is not possible when legal status has not been achieved.

Integration

Healthy integration is the ultimate goal of our work with refugee and immigrant children. Nonprofits play an active role in providing services to children throughout the integration process, and in partnership with other sectors have developed better measures of outcomes for integration than for the domains of stabilization and protection. There are many facets to success in this domain. Essentially, integration is the long-term process by which refugees and migrants establish themselves economically, develop social relationships of various kinds, and participate in the political realm. For most children, success in this domain centers on family relationships, school and preparation for the adult world of work and civic participation. Factors of success such as English language acquisition, academic performance, strong economic status, healthy family dynamics, physical health, and psychosocial adjustment are attained and sustained over time.

However, one of the dilemmas we face is that much of the knowledge about how to achieve success in integration remains at the local level, with the staff of immigrant-serving community-based organizations, refugee service agencies, ethnic mutual assistance organizations, and other nonprofits. The challenge is to surface that learning so that all of the parts of the system can benefit.

Challenges and Recommendations

The protection of refugee and migrant children requires an ongoing commitment to collaboration and coordination across sectors and systems. It is not enough to make information publicly available, although this is a necessary foundation. There must be sustained sponsorship of difficult conversations, locally and nationally, that allow for the exploration of different perspectives and the emergence of new knowledge. Only through such conversations can we sustain a cycle of learning that will work to the benefit of children.

Good information should be made available—in both sophisticated and highly accessible forms—to reach audiences as diverse as academic scholars and researchers, program planners and managers, organizational line staff, teachers, and foster parents. All of the actors require information about new populations of children, data on outcomes in the near and long term, and promising practices.

Ensuring that migrating children are successfully protected also requires that different systems, sectors and professional disciplines are brought together to forge effective working relationships, understand each other's constraints, and commit to specific practices in their own settings that will better support children. Protection is a dynamic process, requiring constant adjustments and change in programs and practices. Nonprofit organizations and the philanthropic community have been a driving force to convene the forums where collaboration, coordination, and information sharing take place. As we deepen our understanding of the challenges and the stakes, it is clear that a more sustained exchange across sectors and disciplines is needed. Let us hope for the sake of children around the world that further opportunities to build our collective knowledge and practice are created.

Notes

1. See Boris, Elizabeth T. and C. Eugene Steurle (eds.) (1999), *Nonprofits and Government: Collaboration and Conflict*, published by the Urban Institute Press in Washington, D.C., for a more thorough description of the distinctive roles of the government and the nonprofit sector, and for the complex relationship between them.
2. Waddell, S. "Societal Learning: Creating Big-Systems Change," *The Systems Thinker*, Vol. 12, No. 10, Dec.-Jan. 2001/2002, Pegasus Communication, p. 1.

17

A Framework for the Care of Unaccompanied Children

Ken Tota
Office of Refugee Resettlement,
U. S. Department of Health & Human Services

This chapter will discuss the efforts of the U.S. government to put together a system of care that considers the interests of each unaccompanied alien child apprehended, and will report on the progress that has been made since the transfer of the Unaccompanied Alien Children's Program from the Immigration and Naturalization Service to the U.S. Department of Health and Human Services, Administration for Children and Families, Office of Refugee Resettlement (ORR).

Numerous countries throughout Europe have begun to tackle the issue of unaccompanied alien children (UAC) and have looked to ORR as a model for providing services to them. On the basis of the efforts in the United States to care for UAC, this chapter outlines a framework of care intended to be helpful for those countries when they make policies about how to safeguard UAC entering their country as refugees or migrants in search of a better life. Many papers on the care of children address the level of care and services; this paper is different in that it will provide key fundamentals that must be in place to facilitate a system that seeks to provide these services.

The U.S. Challenge

Until recently, the United States faced many issues in its care of UAC. In the United States, the Immigration and Nationality Act, which governs the process for determining status, has no provisions that specifically address the care and treatment of UAC or at least distinguish them from adults. Prior to 2003, the Department of Justice (DOJ), Immigration and Naturalization Service (INS), was the lead federal agency entrusted with oversight responsibility for the care of all children apprehended by its Border Patrol officials.

The juvenile program under the DOJ/INS faced a number of issues and was often criticized by many in the advocacy and legal communities. Many of the issues centered on the use of juvenile detention centers, use of shackles, strip searches, poor living conditions, limited recreation, limited access to interpreters, limited access to legal resources, limited mental health resources, the housing of juveniles with adults or criminal offender populations, and inappropriate escorts. The Office of the Inspector General conducted a review of juveniles in INS custody and in September 2001 issued a report that confirmed many of the issues identified.

The INS also encountered criticism because they served as both the custodian and judge, due to their role in determining the status of the child. This was often viewed as a conflict of interest.

Beginnings of a Change

Children without immigration status filed a class action lawsuit in 1985 to challenge the conditions of confinement by the INS. After lengthy litigation in which certain aspects of the lawsuit were won by either the INS or the plaintiffs, the parties resolved the remaining aspects in the settlement now referred to at the *Flores* Settlement Agreement.

ORR follows the requirements of the *Flores* Settlement Agreement (*Jenny Lisette Flores, et al., v. Janet Reno, Attorney General of the United States, et al., Case No. CV 85-4544-RJK (C.D. Cal. 1996)*) which outlines detailed provisions for ensuring that UAC are placed in the least restrictive setting appropriate to the child's age and special needs while in government custody and that they are promptly reunited with family members in the United States when such family is available and when such reunification is appropriate.

The Transfer of Federal Oversight

On March 1, 2003, the Homeland Security Act of 2002 transferred functions under U.S. immigration laws regarding the care and placement of UAC from the commissioner of the INS/DOJ to the director of the ORR within the Department of Health and Human Services.

For many years, ORR has provided services for unaccompanied refugee children as part of its mission to assist refugees in the United States. Caring for UAC is a natural complement to many of ORR's programs and fits squarely within its mission. On the basis of its long history and knowledge of child welfare programs in the United States, ORR has worked to ensure that children are valued, respected, and most of all, protected in the development of this new program.

Through the Homeland Security Act, Congress transferred responsibility to ORR. ORR ensures the basic tenants of the *Flores* Settlement Agreement are incorporated into the overall provision of care thus setting the standard for the treatment of UAC in the United States.

Guiding Principles

- All UAC should be treated with dignity, respect, and special concern for their particular vulnerability.
- ORR personnel shall take into account the unique nature of each child's situation in making placement, case management, and release decisions.
- UAC must be placed in the least restrictive setting appropriate to their age and special needs.
- All UAC shall be provided care and services free from discrimination based on race, religion, national origin, sex, handicap, or political beliefs.

Definition of Unaccompanied Alien Children

Section 462 of the Homeland Security Act (HSA) of 2002 (6 USC §279 (g)(2). defines an unaccompanied alien child as a child

1. who has no lawful immigration status in the United States;
2. who has not attained 18 years of age; and
3. with respect to whom there is no parent or legal guardian in the United States, or there is no parent or legal guardian in the United States available to provide care and physical custody.

This definition of a UAC is often confused with that of a refugee, defined as someone outside his or her country of nationality who is unable or unwilling to return because of persecution or a well-founded fear of persecution, on account of race, religion, nationality, or membership in a particular social group or political opinion.

Profile—Who Are They?

ORR (2008) has reported approximately 7,200 UAC are apprehended every year by Department of Homeland Security (DHS) agents, DHS Border Patrol officers, or other law enforcement agencies and referred for placement. Most come for economic reasons. Often smugglers are paid to help bring them across the border. As of June 2008, approximately 1,429 children were in the care of ORR's Division of Unaccompanied Children's Services (DUCS). The children spend an average of 65 days in ORR care before being released to a sponsor or transported back to their country of origin.

The typical unaccompanied child has traditionally been male, between the ages of 15 and 17, and Central American. But in recent years, there have been increases in the numbers of UAC that are female and of lesser age. UAC come to the United States under a variety of circumstances. Many come to join family members already residing in the United States Others may have no one in the United States and come determined to make a life here on their own. Some UAC apply for asylum, seeking protection from persecution they have already suffered in their home country or fear they will suffer if returned to their country. Others have been smuggled or trafficked into the United States to face forced labor and debt bondage. In 2008, UAC came from the following countries: Honduras (29.7 percent), El Salvador (24 percent), Guatemala (27.7 percent), Mexico (10.7 percent), Ecuador (3.3 percent), and others (4.6 percent).

The U.S. Process

The U.S. structure for apprehension, care, and status determination of UAC is divided among three separate departments. DHS Customs and Border Enforcement is responsible for the apprehension of children and adults at the border and elsewhere, ORR is entrusted with care of UAC once apprehended, and the Department of Justice, Executive Office for Immigration Review (DOJ/EOIR), determines their immigration status. The structure was not always made up of three separate bodies. In 1983 the status determination function was transferred from DOJ/INS to DOJ/EOIR. This was done to remove the function of status determination from an agency whose primary focus was law enforcement and removal. The Homeland Security Act established ORR's role as the third partner in the process whose sole function is care and custody of UAC.

Apprehension

DHS is responsible for the apprehension of all illegal aliens entering the United States When DHS personnel encounter an unaccompanied child, he or she is apprehended and taken to a Border Patrol station. Each local jurisdiction has a DHS Juvenile Coordinator specifically trained to provide intake services and make age determinations for minors. For those in question, dental or bone assessments are conducted. If the agent concludes the child is a minor, the child will be issued a charging document called a Notice to Appear (NTA). The NTA requires the child to appear before an EOIR immigration judge for status determination. Once the NTA is issued, the child is then transferred to ORR for care and custody in one of its shelter care facilities. A determination is made by ORR as to the most appropriate level of care based on initial information received at the time of referral from DHS. Transportation is provided by DHS to the designated ORR shelter facility.

DHS will maintain the minor in the Border Patrol station until transportation can be arranged. During this time children are not to be detained with unrelated adults. Due to the volume of Mexican arrivals, the United States has an agreement with Mexico that allows for their immediate return. INS Border Patrol agents can offer voluntary departure to aliens when they are apprehended at or near the border. It is estimated that close to 80,000 illegal aliens under the age of 18 are stopped at the border each year. Most of these are Mexicans traveling with relatives or other adults and are simply sent back with a "voluntary return" status. This return status means they may attempt to enter again without penalty. Even repeated attempts and apprehensions do not result in penalties or a bar to entry. "Other than Mexicans" (OTMs) are in most instances taken into custody and referred to ORR for placement.

ORR local field staff work closely with their DHS counterparts with regard to placement, release and/or removal. These individuals are specifically trained to address issues related to unaccompanied children. This structure is extremely important for ensuring the best interests of the child. Children require very specific policies and procedures and those who come in contact with them must be fully knowledgeable in this area. ORR field offices are made up of staff with child welfare backgrounds and/or licensed social workers/clinicians.

Placement of Unaccompanied Children

If DHS decides to formally detain the child, the arresting DHS Officer must notify ORR staff to arrange placement and transportation to an appropriate facility. ORR employees take into account the unique nature of the child's situation in making decisions regarding the child's placement.

Most children taken into custody are placed in nonsecure facilities pending their release from custody or return to their country of origin. ORR determines the best possible location by looking at availability of placements, age, gender, immigration status, criminal background, potential for family reunification, mental or physical conditions requiring special services, and location of the child's apprehension.

ORR has developed several care options to meet the specific needs of UAC: Shelter Care; Foster Care (longer-term cases, females, the very young); Group Homes; Staff Secure (minor infractions and escape risks); Secure Facilities (serious criminal backgrounds); and Residential Mental Health Treatment Centers for those with mental health needs.

Facilities

The ORR program closely adheres to the minimum standards for care required by the *Flores* Settlement Agreement. ORR requires licensed shelters to not only comply with the minimum standards and state child welfare laws but to exceed these standards in their provision of child welfare focused care and services.

The *Flores* Settlement Agreement requires each UAC to receive physical care, medical and dental services, needs assessment, intake, educational assessment/services, recreation, reading materials, group counseling, individual counseling, comprehensive orientation, reunification services, visitation, right to privacy, religious access and legal orientation services.

All levels of shelter care facilities must be licensed by their respective states. States have very specific licensing standards to ensure consistency in the level of care, specifying everything from sleeping arrangements to staffing ratios (e.g, eight children per staff during day shifts.) Licensing regulates actual living space, which always includes a separate bed, desk, and closet space or trunk.

Upon arrival into a facility, an intake assessment is conducted, including medical and mental health assessments. Each child is medically screened within 24 hours of arrival. ORR promotes awareness of human trafficking and provides support services to identified victims. Assessment and intake materials have been developed to ensure possible victims are immediately identified and protected. A trafficking intake assessment is conducted upon arrival to determine if they are a victim of trafficking. If it is determined they are a victim, ORR certifies eligibility for services and they become eligible for ORR benefits or the refugee minor foster care program.

Victims are eligible for not only refugee benefits but also Medicaid or food stamps.

Facility staff includes case managers, licensed social workers, and clinicians to ensure the child's welfare is taken into consideration in all aspects of his/her care. Each child is also provided a "Know Your Rights Presentation" by a local attorney (pro bono).

Facilities offer children a structured day with a fully developed educational curriculum. Since the length of stay is relatively short, the curriculum is developed to ensure maximum impact based on the child's educational background and language capability. The main focus is to enhance the child's skills in a manner that could have a positive impact on their life whether they stay in the United States or are returned to their country of origin.

Each facility ensures a high level of services, including access to interpreters, case management, medical health and mental health screening, education, vocational training, recreation, field trips (not offered in staff secure or secure), transportation to court hearings and consulates, individual (once per week) and group (twice per week) counseling, food, clothing, shelter, and access to legal services. All services are provided by culturally sensitive staff.

Security is achieved by a heightened level of staff supervision, low key perimeter fencing, video surveillance, and alarm systems in some locations depending on the level of care necessary. A primary challenge has been balancing the need for security without fostering a detention-type setting. Great effort has been made to create child-friendly, "home-like" atmospheres through the use of large homes as shelters. Buildings are often painted in bright colors, with flags from all the countries of arriving children. Many facilities allow children to create their environment by incorporating the children's artwork into the general landscape and decorating scheme. In some instances, facilities have pools, soccer fields, basketball courts, and child play areas.

A primary focus of the UAC program is to immediately begin the reunification process. Release is determined with priority given to parents, legal guardians, adult relatives (brother, sister, aunt, uncle, or grandparents) or an adult individual designated by the parent or legal guardian. Each sponsor must complete a series of background checks and a family reunification packet, which is used to evaluate the sponsor's ability to fulfill his or her obligations in caring for the child. Sixty-five percent of those in care are released to sponsors.

Lack of permanency is one of the main issues that distinguish the UAC program from traditional child welfare programs. Most UAC do not have a high potential for permanent status in the United States. In many ways the best the program can provide to each child is a skill or knowledge that may benefit the UAC in the long term whether they stay in the United States or are returned to their country of origin.

Legal Standing

UAC generally pursue two options for status in the United States: asylum or Special Immigrant Juvenile Status (SIJS). UAC can be on either the detained (shelter care) or non-detained (released to a sponsor) docket for their immigration hearing; however, the detained docket is generally a much faster process. The third option generally pursued is available to those who have been identified as a victim of trafficking. In such cases, the child may pursue a T-Visa. Other options for relief may be available depending on the particular circumstances.

Since the UAC have been charged by DHS with a violation of immigration law, the NTA charging document is filed by DHS with the immigration court and jurisdiction over the case transfers from DHS to EOIR. In this instance, the UAC will request asylum "defensively" before the immigration judge (rather than "affirmatively" before DHS, U.S. Citizenship and Immigration Services [USCIS]). Gonzalez (2006) stated that DHS/USCIS has jurisdiction over asylum claims filed by applicants present in the United States or seeking admission at a port of entry who have not been placed in removal proceedings. Since the majority of UAC spend a relatively short time in ORR custody and are released to sponsors, they complete their immigration proceedings outside of ORR care. For the period April 1, 2006—March 31, 2007, approximately 16 percent of the 3,548 UAC receiving pro bono services applied for asylum and 22 percent of those were granted asylum.

Children granted asylum become eligible for ORR refugee benefits and state- supported foster care under the ORR Unaccompanied Refugee Minor (URM) Program. UAC who have been denied asylum have the opportunity to appeal the decision, delaying the removal order.

Depending on their circumstances, UAC may apply for SIJS. This is a specific visa grant for those deemed to have no family or alternative for placement and have demonstrated abuse, abandonment, or neglect. With a dependency order from the state, a minor can file a visa petition (I-360 for SIJ visa) with DHS Citizenship and Immigration Services. If approved, the minor then can adjust status to a lawful permanent resident (LPR or "green card"). If this status is granted the child could be eligible for placement in a foster care program.

All children deserve a fair and efficient court proceeding. Legal representation can be helpful to the child in navigating the overall process. ORR recognizes the value of legal orientation and representation for children in its custody and currently funds the Vera Institute of Justice to coordinate pro bono representation. Funding is allocated at the local level to cultivate pro bono representation for children in facilities.

Removal Determination

If the EOIR judge determines the child has no claim to stay in the United States, they are issued a removal order and subject to a five to ten-year ban on returning to the United States. DHS works directly with foreign consulates to ensure a family member is available to receive the child. DHS provides escorted transportation back to the country of origin.

If a child has no one to return to in their country of origin, the issuance of travel documents to return is placed on hold and they may remain in shelter care until they reach adult age. A child who turns 18 may be placed on parole until travel documents are secured. DHS cannot detain indefinitely. In such instances, DHS can parole them on their own recognizance or detain them until their immigration case has been determined. A UAC can ask DHS to depart voluntarily. If DHS approves the voluntary departure the child will not see a judge or go to immigration court. Under certain circumstances, a child may go to court and the immigration judge may allow the UAC to depart voluntarily. A UAC allowed to depart voluntarily concedes removability, but is not barred from returning to the United States. Voluntary departure is considered a form of removal, not a type of relief. Approximately 25 percent of UAC in ORR custody are ordered removed to their countries of origin or choose to depart voluntarily.

Improvements Since the Transfer to ORR

The ORR program has made great strides in expanding upon the tenets of the Homeland Security Act and the *Flores* Settlement Agreement. Underlying principles of *Flores* have provided the foundation for the way the U.S. views this vulnerable population. ORR has built its program around these themes, dramatically changing the overall treatment of UAC entering the United States. At the forefront of all decisions and policies is the desire to consider child's interests and to maintain UAC in the least restrictive setting.

Improvements since the transfer to ORR include specialized shelters, increased use of foster care, enhanced medical and mental health services, focused educational and vocational curriculums, pro bono coordination for legal services, less reliance on secure detention centers, incorporation of detailed policies and procedures, and a greater focus on trained field representatives. Implementation of an ORR field structure has been key to ensuring trained staff are in place to mirror the DHS field structure. The lack of permanency remains a difficult challenge for the program, as well as the fact that most UAC really have a questionable legal basis to stay in the United States, no matter the level of legal assistance or services provided.

A major shift for the program has been the dramatic reduction of UAC placed in secure detention by over 78 percent. As an alternative, shelter, group home and transitional foster care capacity has more than doubled since the program transferred from INS.

Enhancements of mental health services have been a priority since many of the children have been victims of some level of abuse in their journey to the United States. Many are traumatized and have a great deal of anxiety and depression related to their circumstances and separation from their families. The program has contracted with residential therapeutic care facilities for placement of children with mental health needs and has increased the level of clinical staff at each facility to help children deal with such issues through counseling.

Access to legal services has been a major priority since federal funds cannot be directly used to support such services. As a result, the UAC program initiated a contract to cultivate pro bono legal service providers so that more children can be appropriately represented during immigration proceedings.

The program has also placed greater emphasis on the identification of trafficking victims. Shelter staff are now trained to look for signs and conduct trafficking assessments to help identify victims.

These improvements have greatly increased the level of care of UAC within the United States, but they did not just happen on their own. The attention brought to the UAC issue by the advocacy community was instrumental in fostering a bill, sponsored by Senator Diane Feinstein of California, that laid the foundation of what later became Section 462 of the Homeland Security Act, which transferred the program to ORR.

A Framework for Care

The following is a list of principles that ORR has found to facilitate a system of care that strives to meet the child's best interest. It is important to keep the child's interests in the forefront of all decisions.

Care. It is very important that a specific government agency be designated responsibility for the care of unaccompanied children so that there is accountability. Preferably, care should rest with an agency with experience in social services and child welfare oversight. If care is not designated in legislation, an agreement between agencies can be used to lay out roles and responsibilities. This designated agency acts as the temporary guardian of the child. Shelters are supervised by this structure to ensure that they provide the appropriate level of services and that staff are trained and monitored.

Field Structure. It is very important that government agencies designated with apprehension and care (preferably two separate agencies) have a trained field structure. It must have key localized points of contact with language capability. Balancing diverse missions can be difficult to achieve when two very distinct agencies are involved in the process. These agencies must work closely together at the local level to facilitate an appropriate hand off of the child.

Interviews. Those who make first contact should be trained and knowledgeable about procedures for processing and care of an unaccompanied child and have a regional contact/coordinator who has the tools to make appropriate determinations specific to the child. Detailed protocols for intake and placement are necessary. A network of local field coordinators with language capability helps to ensure that action is taken through appropriate channels.

Age Determination. Thorough investigation is needed if a child's age is uncertain. If shelter facility staff are appropriately trained and have medical resources and interpreters, they can conduct the research to include review of pertinent documentation and consultation with consulates and relatives. It is important to always err on the side of caution and assume they are younger. As a last resort, forensic dental examinations, radiographs and bone density tests can be used to determine approximate age.

Detention. If the apprehending agency determines the child is a minor, he or she should be placed in the appropriate setting, such as a shelter with specialized services. Children should always be separated from unrelated adults.

Accommodation. Create a defined structure for referral and placement. Transportation should be a designated responsibility. It is important that shelters are child friendly and have trained staff, educational curriculums, religious access, ongoing monitoring, and language capability. It is a difficult balance maintaining secure shelters without creating detention centers. Open egress could potentially place the child in danger. Trust must be built through a caring staff. Children must feel it is in their best interest to remain, due to the level of services, skills and welcoming atmosphere. Shelters need to instill a sense of community, belonging and hope.

Licensing. Licensing of facilities by a State or local government is important to ensure consistency of services and that staff are appropriately trained and monitored.

Specialization. Specialized facilities in terms of care, duration, cultural sensitivity, language and services available ensure that the child's interests are met.

Trained Staff. Staff caring for UAC should have appropriate child welfare backgrounds and be made up of trained experts, such as social workers, clinicians, attorneys or others in the related field. It is important to facilitate working relationships with other child welfare experts and non-profit organizations so that all decisions are guided and informed. Shelter staff can represent the children's interests (as guardians) if they have a social work background and are appropriately trained. Their expertise can be used to "flesh out" cases in terms of possible asylee/refugee status or identification of a trafficking victim.

Procedures. A definitive manual that outlines policies and procedures is critical to ensure that each party understands their role in the process (field staff and headquarters staff, volunteer agency staff, and facility staff). All decisions should be guided by detailed policies and procedures. Ongoing training of all components ensures understanding, consistency and compliance.

Maturity. The relative maturity of a child can be difficult to determine. An agency must be designated to watch over the child during the determination process and ensure appropriate care. In the United States, ORR in many ways acts as the guardian of the child. It is important that shelters have case managers and field staff who can monitor the process and are trained in this area.

Services. Set a standard for child welfare focused services that ensures each child receives at a minimum physical care, medical and dental services, intake needs assessment, educational assessment / services, recreation, reading materials, group counseling, individual counseling, comprehensive orientation, reunification services, visitation, religious access, right to privacy, and legal orientation services.

Health. Interpreters are the key to a successful health screening. Shelter staff with both language and clinical skills can facilitate the intake process by conducting the initial mental health assessment and services.

Education. Shelters should have a focused school curriculum geared to specific backgrounds and the lack of permanency. Vocational skills are especially important for those who might return to their country of origin.

Legal Orientation. Each child should receive notification about their legal rights, such as a "know your rights" presentation. Attorney representation has proven to enhance a child's potential for asylum status (48 percent granted with attorney vs. 27 percent without).

Status. Creation of a status such as SIJS could provide an option if a child cannot return and asylum is not an option. It is important that children not remain in limbo when their status is in question. Children are especially vulnerable if they have no options for status and take flight to avoid removal.

Database. The creation of a central database is critical for tracking trends, outcomes and a child's overall progression through the process.

Conclusion

The intent of this chapter has been to generate discussion on the issue of UAC by laying out the history, challenges, and basic overarching principles and structures that will facilitate a system of care that keeps the interest of the child at the forefront of all decisions. All systems will be somewhat different, but the challenges in the United States and the improvements to date may provide a basis to build a system that strives to meet those interests.

More fundamental questions as to the issue of permanency and the lack of official status for UAC will continue to present challenges and debate, but at the very least, the framework discussed will help ensure that the basic needs of those most vulnerable are addressed.

Part VI

Looking into the Future

*B*oth the aid organizations providing concrete relief and the scholars studying the situation of young immigrants and refugees are "aiming at a moving target," in the sense that future developments, on the one hand, may in minor and even major ways redefine what it means to be an immigrant and, on the other hand, may more or less dramatically alter the societal structure into which young immigrants are received.

Modern communication and transportation technologies make it increasingly easier for immigrants to stay in touch with their old homelands and to migrate back and forth frequently. This has led some observers to expect the emergence of a new type of transnational migrants. In his contribution, Roger Waldinger, however, points out that the increased potential for transnational contacts does not necessarily mean that today's immigrants actually maintain strong transnational ties. He finds that, even in an age of globalization, immigrants tend to root themselves strongly in their new homeland.

In the concluding chapter, Richard Alba and Hui-shien Tsao discuss the long-standing racial/ethnic hierarchy in the United States that provides the structural framework for the life chances of young immigrants (and everybody else). This hierarchy may be in flux. The authors point out that demographic trends (the relative decline in the number of European Americans who are available to fill the higher occupational echelons) and institutional forces (affirmative action) increase the opportunities for non-European minorities to be upwardly mobile. Yet, as the authors caution, it remains to be seen what the eventual outcome of this period of potential liminality will be.

18

Home Country Farewell: The Withering of Immigrants' "Transnational" Ties

Roger Waldinger

Since the late twentieth century, "globalization" has been the order of the day. With international migration bringing the alien "other" from the third world to the first, and worldwide trade and communications amplifying the feedbacks traveling in the opposite direction, the view that nation-state and society normally converge has waned. Instead, social scientists are looking for new ways to think about the connections between "here" and "there," as evidenced by the interest in the many things called "transnational." Those studying international migration evince particular excitement. Observing that migration produces a plethora of connections spanning "home" and "host" societies, these scholars proclaim the emergence of "transnational communities" (see Glick Schiller, Basch, and Blanc-Szanton 1992, Smith and Guarnizo 1998, Portes 2003, Levitt, DeWind, and Vertovec 2003, and accompanying articles in *International Migration Review*, v. 37, 3).

Evidence of ties that the scholars call "transnational" abounds. While some migrants do move to settle and others settle despite initial plans to the contrary, today's mass international migrations also entail movements of other types, including return migration, repeat migration, and circular migration. Moreover, the movement of so many people across borders generates a huge, subsequent flow of information, goods, and perhaps, most importantly, money, moving back and forth across borders. Changes in technology appear to amplify the impact of these exchanges. Though the simple letter did a remarkably good job of knitting together distant transoceanic contacts during the migrations of the last turn of the century, today's migrants can communicate with the stay-at-homes in any number of ways, doing so with a speed and immediacy that, in the view of many experts, keeps migrants and stayers firmly connected. Likewise, shifts in receiving societies also facilitate the expression of home place attachments. Whereas ties to home and host country were previously seen as mutually exclusive, today's political and ideological environment appears to many scholars as more relaxed: in particular, the shift from melting pot to multiculturalism has legitimized the expression of, and organization around, home country loyalties.

If some scholars look at today's immigration and see home place connectedness as its distinguishing feature, others examine the same reality and find that old country ties inevitably give way to new, just as in the past. As Richard Alba and Victor Nee (2003) have argued in their recent, eloquent defense of assimilation, *Remaking the American Mainstream*, the United States of the turn of the twenty-first century is again demonstrating its extraordinary capacity to dissolve ethnic ties. As Alba and Nee explain, the attenuation of home place connections derives from the dynamics of the migration process itself. Immigration is motivated by the search for a better life, a quest that usually has no inherent relationship to assimilation. Only in some instances is assimilation self-consciously embraced; often, it is precisely the end that the immigrants wish to *avoid*. Nonetheless, the effort to secure a better future—find a better job, a safer neighborhood, a higher-quality school—confronts immigrants with the need to choose between strategies of an "ethnic" or a "mainstream" sort. Insofar as the better future is found in a place where out-group contacts are more plentiful than in the neighborhoods or workplaces where the newcomers begin, the new Americans are likely to select "mainstream strategies"—and thereby progress toward assimilation, whether wanted or not. In the process, ties to the countries from which the immigrants stem inevitably lose salience; even though connections and connectedness may not totally wither, they lose practical relevance, retaining mainly symbolic value.

Whether indeed it is transnationalism or assimilation that best illuminates today's unfolding reality, the crucial question has to do with change over time. Since home country connectedness is an integral part of the immigrant phenomenon, one expects recent immigrants to be maintaining close and regular contacts with the people they left behind. Likewise, as contemporary immigration is at once massive and continuing, it is no surprise to find that it has generated the infrastructure needed to facilitate "here-there" contacts and exchanges of all sorts, and thereby also reducing their costs.

But what of the long term? Right from the beginning, skeptics of the transnationalist perspective contended that it was all a matter of spurious correlation: yes, there are lots of "here-there" flows, but control for time spent in the United States, and the degree of home country connectedness, dramatically declines. While debate over the question continues unabated, scholarly attention increasingly focuses on the experiences of the immigrants' children—whose behavior will determine whether home country attachments will exercise an enduring influence on the American scene.

On the whole, the essays assembled in Peggy Levitt and Mary Waters's pioneering collection, *The Changing Face of Home: Transnational Lives of the Second Generation* (2002), provided little supporting evidence for the strong versions of transnationalism, according to which a substantial share of today's immigrants live in social fields effectively spanning host and home societies. While the book's contributors clearly show that second generation home country connectedness persists, regular home country involvements engage a relatively small portion of today's immigrant offspring, for whom the "here-there" tie is at best of modest salience. To be sure, as Levitt and Waters note, even selective transnational engagements can add up. But when as ardent a scholarly transnationalist as Peggy Levitt is forced to conclude that "there may still be a small, but important number who continue to contribute to the political, economic, and social life of their ancestral homelands" (Levitt 2002: 145), one realizes that the phenomenon in question bears little relationship to transnational*ism* as condition of being—which is what this particular suffix means.

This disappointing conclusion notwithstanding, the debate is hardly over; rather, it has just begun. To begin with, the research is still at a very early stage. As argued by Alejandro Portes (2003), most of the relevant research on "immigrant transnationalism" has been conducted using qualitative techniques, and most of the qualitative studies suffer from the deficiency of having sampled on the dependent variable, illuminating the incidence of second generation home country involvement, without being able to tell us much about its prevalence. To some extent, scholarly appreciation of the range and correlates of immigrant transnationalism has improved, thanks to results generated by the "Comparative Immigration and Entrepreneurship Project," a survey of Colombian, Dominican, and Salvadoran immigrants undertaken by Alejandro Portes and his associates. However, caution is advised when generalizing beyond the three relatively small, sociologically distinctive populations in question. Not only are these particular groups unlikely to be representative of the Latin American – origin population living in the United States, the samples, themselves, are unlikely to be representative of the three specific populations in question, as the survey conducted by the Comparative

Entrepreneurship and Immigration Project entailed a significant referral element. As shown by Jose Itzigsohn and Silvia Giorguli Saucedo (2002), levels of transnationalism are generally much higher among the sample's referral component, as compared with those randomly surveyed.[1]

As it happens, we can do better: there are now a number of large, representative samples containing data that allow us to explore the importance of home country versus sending country involvements of the new Americans. While none of those surveys provide information on the behavior and attachments of the children of immigrants, they do allow us to examine a reasonable proxy group, namely, those immigrants who arrived in the United States as children, under the age of 13. This group, commonly labeled as the "1.5. generation," is unlikely to behave just like the members of the "true" second generation, that is to say, the U.S.-born children of foreign born. Nonetheless, as many other studies have shown, the 1.5 generation bears a close resemblance to the "true" second generation on a broad variety of indicators—a reason for us to assume that the same pattern will hold when we shift the focus to look at the relative importance of sending and receiving country attachments.

This chapter makes use of the public use data sets from four large-scale surveys: the 1999 *Washington Post*/Kaiser Family Foundation National Survey of Latinos, the 2001 Pilot National Asian American Political Survey (PNAAPS), the 2002 Pew Hispanic Center National Survey of Latinos, and the 2004 Pew Hispanic Center Survey of Mexican Migrants. The Kaiser and the 2002 Pew Hispanic surveys were nationally representative ones, using highly stratified, random digit dialing samples to obtain foreign-born respondents. The PNAAPS involved a semi-random sample of Asian American households in the metropolitan areas of Los Angeles, New York, Honolulu, San Francisco, and Chicago. The 2004 Survey of Mexican Migrants was conducted among Mexican migrants visiting consular offices in the United States, as they were applying for a *matricula consular*, an identity card issued by Mexican diplomatic missions. Unlike the first three surveys, the latter was a purposive survey, especially likely to capture undocumented immigrants, seeking identity documents from the Mexican government. While caution needs to be used in generalizing from this sample to the broader Latino population, this is precisely the group most likely to maintain close home country connections, which is why any possible differences between first and 1.5 generation respondents are of particular interest.

The use of multiple surveys yields both advantages and disadvantages. On the positive side, consistency—defined not so much by point estimate as by modal pattern—lends confidence in the reliability of results. On the downside, coverage is not comprehensive, with the full range of relevant topics not covered by any of the surveys. As a totality, however, the surveys do cover many of the issues in question, allowing us to take a broad, multidimensional look at the phenomenon.

In this chapter, I use these data to compare 1.5 generation immigrants, as defined earlier, with first generation immigrants, defined as persons born *and* raised abroad. The chapter provides new evidence, both on the linkages that tie first and 1.5 generation immigrants to the countries from which they have come *and* on the new connections to the United States that are developing as the immigrants put down roots. I break the complex of "here-there" ties into several components:

1. *Home country engagements, including travel, remittances, and home country politics*
2. *Identity and attachment*
3. *U.S. political participation*

Findings

Cross-border exchanges and ties: While the mass migrations of the turn of the last century produced a huge traffic of correspondence, coming and going across the Atlantic, today's mass migration occurs at a time when communication is instantaneous. Changes in technology have made it possible for migrants to communicate with their friends and relatives living back home while ever intensifying competition among the providers of telecommunication have driven costs down. As a result, almost any immigrant can afford to spend a few hours per month communicating with relatives still living back home.

Although there can be no doubt regarding the *potential* for regular cross-border communication, the surveys suggest that the reality falls a good deal short of the possibilities that current technology allows. Among the Asian immigrants polled by the PNAAPS, roughly two-thirds were in regular contact (once a month or more frequently) with persons living in the home country. The Pew survey of Mexican migrants found even higher, though again not universal, levels of regular home country communication, a pattern to be expected considering the fact that this survey sampled a much more recently arrived population. Nonetheless, in both cases the 1.5 generation respondents were significantly less likely to engage in regular home country communication than were the first generation respondents. Not only is the difference large, but the proportion falling out of regular contact is sizeable, reaching the one-third level among the Mexican 1.5ers and exceeding the half-way mark among the 1.5ers polled by the PNAAPS, as can be seen from Table 18.1 (table 18.1).

Travel is yet another way by which immigrants are said to maintain their regular home country connections. As shown by both the PNAAPS and the Pew National Survey of Latinos, the great majority of immigrants are likely to have taken at least one trip back home after migration to the United States. That rates of return travel should converge among Latino and Asian immigrants is striking: on the one hand, Asian immigrants have greater resources and are more likely to possess legal status, both of which are characteristics likely to be associated with return travel; on the other hand, distances involved are much smaller for the Latinos, which implies that costs were correspondingly reduced. Most importantly, perhaps, generation makes little difference in either group, with the great majority of both first and 1.5 generation respondents likely to report having taken at least one trip to their former home country.

Quite a different pattern, however, emerges when the focus shifts to the sending of remittances. A majority of immigrants appear *not* to send home remittances. However, the proportion engaged in remittance sending involves a large minority; among Latino immigrants that fraction may come close to one-half. Given the size of the denominator, one can readily understand the interest that remittance sending generates among home country governments and international development agencies.

The high and growing levels of remittance dependency among many of the emigration-sending countries in the Western Hemisphere also explain why governments should be concerned about the prospect that the flow of remittances may one day falter. The results from the Kaiser and 2002 Pew surveys indicate that those worries are indeed well founded, with members of the 1.5 generation significantly less likely than members of the first generation to send remittances to their former home countries. If roughly one-half of all Latino first generation immigrants send home remittances, it appears that only one-quarter to one-fifth of the 1.5 generation do the same. Depending on the perspective, the latter may be considered high or low. Sending remittances is a tangible expression of solidarity and connection with people living in another country, and thus evidence that assimilation is far from complete. On the other hand, the fact that most 1.5 generation immigrants have fallen out of the group of remittance senders testifies to the long-term power of the forces cutting ties to former home countries. In this respect, findings from the Survey of Mexican Migrants are telling. While remittance levels are high among both generational categories, the disparity between the two is very sizeable, with 1.5 generation respondents only about 60 percent as likely as their first generation counterparts to send home remittances.

Table 18.1 Cross-border engagements

	Routine cross-border exchanges			Home country politics		
	Sends remittances	One trip or more	Calls home country less than 1x mo	Voting	Any participation	Has election credential	Belongs to hometown association
Kaiser 1999							
all	41%			25%			
first generation	45%			28%			
1.5 generation	**24%**			**4%**			
PNAAPS 2001							
all		79%	61%		6%		
first generation		79%	63%		5%		
1.5 generation		81%	**41%**		11%		
Pew 2002							
all	47%	72%		15%			
first generation	50%	72%		16%			
1.5 generation	**21%**	69%		10%			
Pew Mexican Migrant Survey 2005							
all	79%		85%			42%	14%
first generation	81%		87%			44%	14%
1.5 generation	**48%**		**65%**			**16%**	16%

Note: Tests are for significance of differences between first and 1.5 generations; bold = p < .01.

Social and economic ties are probably the most important of the home country connections maintained by immigrants; nonetheless, participation in home country politics has been an immigrant perennial. The clamor for home country voting rights is clearly growing, at least among some immigrants with the resources needed to make their voice heard. Likewise, home country political organizations are extending their infrastructure to the United States, on the lookout for funders, but also responding to the emigrants' persistent interest in home country politics.

From the surveys emerges the following picture. At the high end of the spectrum, the Kaiser survey suggests that a quarter of Latino immigrants have voted in their home country at least once since moving to the United Sates; at the lower end, the PNAAPS indicates that barely 6 percent of Asian immigrants have had any involvement in home country elections since movement to the United States (in part probably due to one-party rule in several large Asian countries). Among Asian immigrants, the 1.5 generation appears to have experienced somewhat higher levels of participation than has the first generation, though in either case, it is a very small minority. Among Latinos, the 1.5 generation reveals much lower levels of home country involvement than does the first.

As other studies have shown, interest in home country voting is often limited to an immigrant elite, who possesses the resources needed to maintain active involvement in home country matters. Pressures from immigrant elites yield responses from sending country officials eager to maintain the flow of remittances and other investments, and also to convert the immigrants into ethnic lobbyists. Nonetheless, home country leaders are often wary of demands for political participation, for ethical, self-interested, as well as practical reasons. Among the latter, a key concern involves the means of verifying eligibility for participation. As suggested by the Pew Survey of Mexican Migrants, that worry is indeed well grounded: only 42 percent of all respondents claimed to have a Mexican voter credential in their possession, which suggests that even among these recently arrived immigrants, who are most likely to retain strong interest in developments back home, the potential for participation is unlikely even to reach the halfway mark. Among the 1.5ers, however, possession of the credential is much rarer—an indicator that, along with results from the Pew and Kaiser surveys, strongly points to detachment from home country politics among this group of immigrants who moved to the United States as young people.

Yet another indicator of homeland-oriented interest is the formation of hometown associations. The 2005 Pew Mexican Migration Survey asked respondents whether they were involved in a sports, civic association, or club with people from their home country or hometown. Roughly one out of every seven respondents answered "yes"; 1.5ers were as likely as their first generation respondents to be involved in associations of this sort. These responses are consistent with the results of other research, all of which indicate that civic activity undertaken by immigrants in the United States but oriented toward hometowns or homelands engages a small proportion of the immigrant population. Considering the underlying denominator, however, the absolute number of persons so involved is large; given the resources available to these immigrant activists, moreover, the impact that any collective activity *might* yield on hometowns or home societies is far from trivial. As regards the relative importance of transnational versus territorializing processes, however, these data further underline the attenuation of home country ties.

National attachment and identity: Clearly, immigrants continue to maintain connections to their old home country, albeit at levels that vary greatly, depending on the dimension, and with 1.5 generation immigrants generally more detached than the first. But even while money and people continue to move back and forth between receiving and sending countries, settlement proceeds ahead. The myth of return is a common component of the immigrant experience; nonetheless, most immigrants realize that migration to the United States involves a permanent change. The surveys suggest that among Latinos, roughly two-thirds plan to stay in the United States for good; among Asian Americans, the level is close to three-quarters. Whereas many members of both the first and 1.5 generations expect that residence in the United States will be permanent, plans to stay for good are endorsed by considerably higher levels of the 1.5 generation, regardless of survey and/or ethnic background. Once again, as can be seen from Table 18.2, results from the 2005 Survey of Mexican Migrants are telling: even among this group of relatively recent arrivals, many lacking legal status, expectations of return are low (table 18.2).

Like many immigrants before them, today's new Americans are seeking to combine the new and old, even as they put down roots in the United States. For example, two-thirds of both first and 1.5 generation respondents, when polled by the Pew Hispanic Center in 2002, said that their core self-identity was linked to their country of origin. On the other hand, the majority of first generation respondents continue to think of the home country as their real home, but barely one-third of the 1.5 generation respondents chose the same response. Although limited, the data on identification and attachment suggest that the immigrants are "in-between," shifting loyalties and allegiances toward the United States, even as they are trying to remain true to the people and places they have left behind.

U.S. Political Participation: As the burgeoning literature on "immigrant transnationalism" has shown, connections linking sending and receiving

Table 18.2 National attachment and identity: The United States versus home country

	will stay in the United States	real home is home country	self-identity: home country
Kaiser 1999			
all			
first generation			
1.5 generation			
PNAAPS 2001			
all	78%		
first generation	94%		
1.5 generation	77%		
Pew 2002			
all	65%	61%	68%
first generation	63%	65%	69%
1.5 generation	83%	**36%**	63%
Pew Mexican Migrant Survey 2005			
all	64%		
first generation	62%		
1.5 generation	86%		

Note: Tests are for significance of differences between first and 1.5 generations; bold= p < .01.

countries are easy to find, in large measure because connectedness is an inherent aspect of the migration phenomenon. For that reason, the frequently repeated claim that migrants "*may* continue to participate in the daily life of the society from which they emigrated but which they did not abandon" (Glick-Schiller 1999: 94, emphasis added), amounts to little more than the null hypothesis. The crucial issue, rather, as correctly noted by Peggy Levitt and Nina Glick-Schiller (2003: 1003), concerns "the relative importance of nationally restricted and transnational social fields" (table 18.3).

While this is a question to which little of the literature has yet responded, a comparison of migrant political participation in receiving and sending countries yields insight into the relative importance of territorial versus trans-state factors. As noted earlier, relatively few migrants engage in home state politics. One might suspect that immigrants might be similarly detached from politics in their adopted land. In contrast to the last era of mass migration, political machines are no longer actively mobilizing immigrant voters, and weakened party structures provide no

Table 18.3 U.S. political participation (naturalized citizens only)

	Registered voter	Voting	
		Ever voted	Voted last presidential election
Kaiser 1999			
all	74%	70	53%
first generation	75%	68	49%
1.5 generation	77%	74	**64%**
PNAAPS 2001			
all			48%*
first generation			
1.5 generation			
Pew 2002			
all	82%	72%	
first generation	84%	73%	
1.5 generation	75%	67%	
Pew Mexican Migrant Survey 2005			
all			
first generation			
1.5 generation			

Note: Tests are for significance of differences between first and 1.5 generations; bold= $p < .01$.
* sample of citizens too small for further disaggregations.

ready substitute. Citizenship requirements are steeper than before, with the result that naturalization rates have declined; furthermore, a large portion of the immigrant population lacks legal status, and thus is disconnected from American politics on a long-term basis.

Notwithstanding all the trends that might foster detachment from the U.S. political system, the available data show that, among naturalized immigrants, receiving society political participation is far more common than sending society participation. Among immigrant Latinos, the surveys indicate that three-quarters or more of the naturalized citizens report that they are registered to vote. Not all registered voters go to the polls, but the Pew and Kaiser surveys suggest that, at least occasionally, the great majority do. By contrast, the Kaiser survey, which asked about voting in the last presidential election, indicates that regular electoral participation is less common.

Generational differences, however, are slight, though survey results are not fully convergent. In general, it seems fair to say that 1.5 generation respondents manifest levels of participation that are comparable to those of first generation respondents, a convergence that is remarkable in light of the facts that (a) the former are younger than the latter and (b) political participation generally grows with age. Thus, contrary to the claims that transnational migration involves "the process of simultaneous incorporation into immigrants' states of origin and settlement" (Glick-Schiller 1999: 286), these data point to a different pattern. Rather than immigrants maintaining more or less constant links to the home society while establishing comparable connections to the host society, home country involvements actually falter, proving greatly outpaced by levels of host society participation.

Conclusion

Like the turn of the twentieth, the turn of the twenty-first century is an age of mass migration, with large numbers of people seeking to move from poorer to richer countries. To get from one place to another, the migrants make use of the one resource on which they can almost always count—namely, support from one another, which is why social connections between veterans and prospective movers lubricate the migration process.

In the mythology of the classic countries of immigration, the newcomers are arriving in order to build a life in the new land. In reality, it is often not the case, as the migrants instead want to take advantage of the gap between richer and poorer places to accumulate resources intended to be used upon return back home. Although some migrants eventually act on their plans, for others return turns out to be a myth, as roots get established in the country of arrival, whether wanted or not.

Considering the centrality of migrant networks, the myriad of migration strategies, and the uncertain, transitional nature of the migration process, connections linking origin and destination places are ubiquitous, no less characteristic of today's age of mass migration than of that of the past. But if the ongoing advent of new immigrants keeps here-there connections refreshed, the long-term tendency goes toward the attenuation of those contacts, as relevant social ties and loyalties get transplanted from old to new homes (see Waldinger and Fitzgerald 2004). Although immigrant offspring—whether born or raised in the host society—are likely to retain affection for the old home country and at least some home country connections, their lives and identities get firmly rooted in the nation-state society in which they are making their lives.

That generalization holds even for the young people studied in this chapter, who would be the ideal candidates to retain home country attachments. As shown, home country connectedness is limited. While most immigrants view the United States as home, the 1.5ers have put down deeper roots and have steadily cut back their involvements in, and ties to, the countries where they were born. Although we still know too little about participation in U.S. institutions, the available indicators tell us that, for all immigrants, host country engagement is a good deal higher than the forms of home country involvement identified by the various surveys we have used.

Although the long-term trends are clearly toward settlement in and commitment to the United States, this chapter also shows that immigrants are in a process of transition. Some immigrants do sever all ties to their old home countries; however, a significant minority continues to engage in a variety of home country – oriented activities. But the best way to characterize the immigrants' "here-there" connection is to describe them as still "in-between": the immigrants are moving along a trajectory of shifting loyalties and allegiances toward the United States, even as they are trying to remain true to the people and places that they have left behind. Persons impatient with immigrants' progress might insist these home country connections be cut, and the sooner the better. The better approach is simply to let developments follow their own course: with time, the immigrants will find their own way to go from "there" to "here."

Notes

1. Itzigsohn and Saucedo explain that "about a third of the sample (37.7 percent) was selected through referrals and snowball chains with different points of entry. The reason for the purposive selection of one third of the sample was that the principle focus of the survey was to study transnational practices. The referral section of the sample attempted to reach people who engage in transnational practices to insure that there were enough cases of transnational migrants in the sample" (Itzigsohn and Saucedo 2002: 769). As shown in Table 3 of the article (p. 776), the level of transnational practices was consistently higher within the referral than within the block sample: for example, 46 percent of the referral sample participated occasionally or regularly in a hometown association versus 14 percent of the random sample, 45 percent of the referral sample sent money for projects in a hometown versus 14 percent of the random sample, 31 percent of the referral sample participated in local sports clubs linked to the home country versus 10 percent of the random sample, and 60 percent of the referral sample participated in charity organizations linked to the home country

versus 15 percent of the random sample. As shown in Appendix Table 2, the two samples also differed along other variables, likely to influence the outcomes: for example, citizens comprised 50 percent of the referral sample but only 24 percent of the random sample, highly educated persons (with 13 or more years of schooling) made up 63 percent of the referral sample but only 26 percent of the block sample, and men were 70 percent of the referral sample but only 45 percent of the random sample. Thus, Portes's description of the traits of the "transnationals" (as quoted in the previous paragraph) seems to largely reflect the characteristics of the members of the referral sample.

19

Is There a Looming Period of Liminality for Race and Ethnicity in the United States?

Richard Alba and Hui-shien Tsao

Social scientists take for granted that racial and ethnic origins play a critical, though hardly the exclusive, role in determining the life chances of Americans, including where they live, how much education they get, what kinds of jobs they do, and whom they marry. An enormous literature establishes that this assumption is generally warranted. For some of the most salient racial divides in the United States, such as that between blacks and whites, that literature also demonstrates that numerous differences in life chances have remained stable for decades. One particularly significant instance concerns residential life chances, which not only involve who the neighbors will be, but what the "quality" of the neighborhood is—reflected in the risk of criminal victimization, the adequacy of the schools, or other ways. The research on residential segregation reveals the stability of black-white differences in these respects for at least half a century (Massey and Denton 1993, Logan, Stults and Farley 2004). In this sense, a crystallized racial/ethnic order could be said to exist in the United States, with whites occupying the top position and African Americans at the bottom, with others somewhere in between.

There are sound reasons to think that this order influences the chances for success of the second generation. The segmented assimilation theory about the incorporation of new immigrant groups asserts as much (Portes and Zhou 1993), and empirical research sustains the view that the children of immigrants find themselves in a society where their options are constrained by the way other Americans view their racial/ethnic membership, even when this does not coincide with the perceptions and beliefs their parents have brought from their societies of origin (Waters 1999, Kasinitz et al. 2008).

Yet, looking at the past, we also know that the emergence of a massive second generation can unsettle racial and ethnic relations and lead to a reshuffling of the hierarchical order among groups. Enormous literatures, on assimilation and on whiteness, address these changes in the past. They suggest that we ought at least to entertain the notion of some degree of racial and ethnic change in the next few decades, with the arrival of a large and upwardly striving second generation on the scene (see also Alba and Nee 2003, Bean and Stevens 2003). This is what we would like to do in this chapter on the basis of some results about racial and ethnic shifts among the incumbents of highly ranking occupations. We will argue that the shifts already evident, resulting probably from demography and affirmative action, combined with those shifts that can be anticipated as a result of the departure of the baby-boom cohorts of whites from the labor market, indicate an opportunity for important changes to the racial/ethnic order in the next quarter century (for a fuller statement, see Alba 2009). If this opportunity is realized, something that is certainly not assured, it will likely usher in a period of unsettlement, when established assumptions about relations between individuals derived from their categorical memberships lose their certainty. This is the justification for the word "liminality" in the title, for it refers to a period of transition, when previously fixed identities are in flux. In the final part of this chapter, we will consider, on the basis of our knowledge about the incorporation of earlier immigrant groups, some contingencies that will affect whether this opportunity is realized or not.

Our Imperfect Understanding of the Past and Why It Matters

Racial and ethnic change was a major aspect of the assimilation of European groups during the first six decades of the twentieth century (through, say, the election of John F. Kennedy as the first Catholic president). Irish Catholics and southern and eastern European Catholics, Orthodox Christians, and Jews were initially viewed as racial outsiders by native white American Protestants, and occupied an in-between position in U.S. labor and housing markets (Foner 2000, Perlmann 2005, Roediger 2005, Foner and Alba 2006, see Foner and Ueda, both in this volume). The derogatory language that was routinely applied to southern and eastern European immigrants and their families—for example, "hunkie" and "guinea"—betrayed their problematic position. Nowhere was this better registered than in the "guinea" slur for Italians, which ultimately derives from the history of slavery, as it refers to the West African coast as well as to the black bondsmen who came from there (Roediger 2005: 37, see also Alba 1985). "Scientific" racism, alleging the inferiority of the immigrants, provided a major rationale for the restrictive immigration legislation of the 1920s and its national-origin quotas (to say nothing of the complete exclusion of Asian immigrants [Ngai 2004]).

Yet, during the second half of the twentieth century, it became apparent that the racial position of these once-despised groups had shifted fundamentally. In the contemporary parlance, they had been fully incorporated as whites. Without question, they caught up over time to other whites in terms of socioeconomic position and integrated with them in suburbs and through intermarriage (Lieberson and Waters 1988, Alba 1995, Perlmann and Waldinger 1997, Alba and Nee 2003, Perlmann 2005). We do not yet possess a completely satisfactory account of how this racial/ethnic uplift occurred. Current theories among historians place the emphasis on the initial legal position of white ethnics, who, unlike Asian immigrants, could naturalize and thereby attain a modicum of political influence and who were not barred, as were nonwhites in many states, from intermarriage with other whites (Jacobson 1998, Roediger 2005). Also recognized as important is the intermediate position of the ethnics in the labor market, which translated into greater access to union membership, significant in an era when unions still monopolized many skilled jobs. Considerable weight is also laid on the consequences of New and Fair Deal policies, as entailed in the legislation and implementation of such innovations as the Federal Housing Authority, the Social Security Act, and the GI Bill, which in their totality gave advantages to ethnic whites in a variety of domains that were withheld from the great majority of African Americans and other nonwhites. The political scientist Ira Katznelson (2005) has characterized the relevant laws and policies as "affirmative action" for whites.

This historical account, however, slights something that we want to emphasize here: the occurrence in mid-century America of non-zero-sum mobility, generated in this case by massive changes in socioeconomic structures. Such mobility does not require downward mobility by members of more privileged groups for upward mobility of the less privileged to occur. The structural changes enabling non-zero-sum mobility are indicated in the transformation of the college and university system, which in a period of just a quarter century (1940–1965) expanded several times over and thus accommodated many more students than before; the great expansion in the middle and upper reaches of the occupational system, creating room for newer groups to move up; and the drastic reorganization of residential space, characterized by the emergence of many new suburban communities where white families of diverse ethnic origins could buy homes and mix. One way of summarizing these changes is to say that they occurred through structurally generated mobility that brought second and third generation ethnics into positions of parity with other white Americans at work and outside of it, because of educational and occupational advance, residential mobility, and intermarriage.

We will claim here that non-zero-sum mobility has a special significance for racial/ethnic boundary change. It implies that upward mobility is less likely to be accompanied by an intensification of competition along ethnic and racial lines, and the lower temperature of competition allows for the relaxation of boundaries. Note, in particular, that for the white ethnics this relaxation is not only a matter of working alongside previously more privileged whites, but also of achieving greater social proximity to other whites through residential and other changes.

On the basis of this brief analysis, the next question obviously is, will non-zero-sum mobility occur in the future in a way that affects the prospects for the second and third generations issuing from contemporary immigration? Might such mobility also affect African Americans? There is pessimism about the possibilities for mobility today because of economic structural changes, and it undergirds the original formulation of segmented assimilation theory (Gans 1992, Portes and Zhou 1993). However, this pessimism overlooks the likelihood of mobility occurring as the number of European Americans available to take good jobs declines not only relatively, but even absolutely. This decline is predictable from the demographic shifts of the U.S. population. The pessimistic view also overlooks the potential impact of affirmative action—especially in higher education (Bowen and Bok 1998, Skrentny 2002, Massey et al. 2003)—and thus the likely rise in the number of minority group members who are positioned to take good jobs. These demographic and institutional forces could mean, then, mobility for some individuals who are now regarded as members of non-European minorities. The growth of the middle-class portions of these groups could weaken ethnic and racial divisions, if mobility is also associated with a narrowing of the social distance from whites.

Some Contemporary Evidence of Racial/Ethnic Shift

Shifts are visible in the recruitment into good jobs in the American economy. To show them, we rank-order occupations by the average remuneration received by their full-time incumbents. We then slice this hierarchy in ways that account for different fractions of the full-time labor force, for example, the best-paid occupations needed to account for 10 percent of the labor force, or the top decile, and the top quartile as defined equivalently. It should be noted that individuals are presented here according to their (detailed) occupation and its average pay for full-time workers, not according to their own pay. This prevents the classification from suffering from bias against the achievements of younger workers, who, though they may be pursuing well-paid occupations, are often paid below the average. The data come from the 2000 Census and are supplemented with data from the 2005 American Community Survey.

Table 19.1 shows the racial, ethnic, and nativity characteristics of the incumbents of top jobs in the top decile and quartile by age group (birth cohort) according to the 2000 Census. This table demonstrates unequivocally that the racial/ethnic backgrounds of the holders of the best jobs in the U.S. labor force are changing. The reasons for the shifts are subject to speculation, but it is plausible to think that demographic changes in the population and affirmative action figure prominently among them. In the oldest age group, individuals aged 55–64 in 2000, 84 percent of the incumbents of the best jobs, whether defined as those of the top quartile or decile, were native-born non-Hispanic whites. This fraction decreases very slightly in the next oldest age group (45–54 in 2000), but is lowered more noticeably with each younger group. Among those aged 25–34 in 2000, only 74 percent of the top quartile jobs and 72 percent of the top decile are native-born whites. Indeed, 22–23 percent of these top jobs are held in this age group by nonwhites and Hispanics. This fraction has doubled between the oldest and youngest cohorts.

Table 19.2 shows the equivalent data for the same birth cohorts as updated in the 2005 American Community Survey; to them we add the data for relatively new entrants to the labor force, aged 25–29 in that year. The table emphasizes even more strongly the split between the older cohorts and the younger ones. While a shift in the direction of greater minority representation begins with the cohort then aged 40–49, it becomes very pronounced in the two youngest cohorts, in the ages below 40. In fact, it appears to be occurring not only across cohorts but also within them over time: a comparison of the the 30–39 age group of 2005 to the same birth cohort in the 2000 data reveals that the increasing representation of nonwhites and Hispanics has continued in the first years of the twenty-first century; consequently, the fraction of top quartile jobs occupied by U.S.-born non-Hispanic whites has declined from 74 to 70 percent. The levels of minority representation in the youngest cohort, aged 25–29, are the same as those in the 30–39-year-old group.

Both tables, but especially the second, demonstrate that the gains in minority representation in top occupations are going only partly to the U.S. born, including the second generation. To be sure, in table 19.2, the

Table 19.1 The racial, ethnic, and nativity composition (%) of the best-paid occupations by age group, 2000

	NB non-Hisp whites	FB non-Hisp whites	NB non-Hisp blacks	FB non-Hisp blacks	NB Hisp	FB Hisp	NB Asians	FB Asians	NB others	FB others	N (1000s)
Ages 55–64											
Top quartile	83.8	4.8	3.6	0.5	1.3	1.4	0.4	2.9	1.0	0.3	2,555.5
											100.0%
Top decile	83.9	5.5	2.6	0.3	1.1	1.4	0.5	3.6	0.9	0.3	1,213.1
											100.0%
Ages 45–54											
Top quartile	82.3	4.1	4.7	0.6	1.8	1.5	0.6	3.0	1.2	0.4	6,296.3
											100.0%
Top decile	82.5	4.7	3.8	0.4	1.6	1.4	0.7	3.5	1.1	0.4	3,029.0
											100.0%
Ages 35–44											
Top quartile	78.1	4.6	5.1	0.9	2.8	2.1	0.7	4.1	1.2	0.5	7,403.2
											100.0%
Top decile	77.8	5.4	4.3	0.7	2.4	1.9	0.9	5.1	1.1	0.5	3,705.6
											100.0%
Ages 25–34											
Top quartile	73.9	4.2	5.6	0.7	4.1	2.3	1.2	5.9	1.5	0.6	5,249.0
											100.0%
Top decile	72.0	5.0	4.8	0.7	3.4	2.1	1.6	8.4	1.4	0.7	2,525.9
											100.0%

Note: The top quartile encompasses the best-paid occupations sufficient to account for a quartile of the full-time labor force; the top decile is defined equivalently. Aside from "others," the racial categories include only individuals who report unmixed racial backgrounds; those with mixed racial backgrounds are placed in the "other" group.

Table 19.2 The racial, ethnic, and nativity composition (%) of the best-paid occupations by age group, 2005

	NB non-Hisp whites	FB non-Hisp whites	NB non-Hisp blacks	FB non-Hisp blacks	NB Hisp	FB Hisp	NB Asians	FB Asians	NB others	FB others
Ages 60–69										
Top quartile	82.5	4.6	3.3	0.6	2.0	1.6	0.5	3.8	1.0	0.1
Top decile	82.7	5.2	2.5	0.4	1.4	1.6	0.5	4.5	0.9	0.2
Ages 50–59										
Top quartile	81.7	3.8	4.6	0.8	2.4	1.5	0.7	3.4	1.0	0.2
Top decile	82.3	4.8	3.5	0.6	1.9	1.4	0.8	3.8	0.8	0.1
Ages 40–49										
Top quartile	77.0	3.8	5.7	1.1	3.4	2.2	0.9	4.7	1.2	0.2
Top decile	77.0	4.8	4.5	0.7	2.8	2.0	1.1	5.9	1.2	0.2
Ages 30–39										
Top quartile	70.0	4.0	6.7	1.0	5.0	2.8	1.4	7.5	1.4	0.2
Top decile	67.9	5.1	5.5	0.8	4.2	2.3	1.9	10.7	1.2	0.3
Ages 25–29										
Top quartile	70.2	3.0	6.5	0.8	6.5	2.5	2.6	6.1	1.6	0.3
Top decile	68.1	4.0	5.4	0.6	5.2	2.3	3.4	9.1	1.6	0.4

representation of U.S.-born Hispanics has tripled between the oldest and youngest cohorts and that of U.S.-born blacks has doubled. Nevertheless, immigrants, especially from Asia, are also increasing their share. Overall, for instance, the foreign born account for 16 percent of top decile jobs among the 25–29-year-olds of 2005, with the Asian foreign born alone taking up 9 percent. We have not yet determined how many of these foreign-born individuals are members of the 1.5 generation, who have grown up in the United States in immigrant households. Almost certainly, though, a hefty proportion is made up of individuals who arrived in the United States as adults, either during their student years or afterward. The large share of top jobs filled by immigrants reveals an alternative to the recruitment of U.S.-born minorities that could compensate for any decline in the number of non-Hispanic whites available for these positions.

In any event, equality of chances to occupy the best jobs has not been attained. Whites remain very privileged. Thus, non-Hispanic blacks represent 12–13 percent of the 25–29 age group in the 2005 American Community Survey (which does not include, it should be noted, the incarcerated population), but they constitute only about half that fraction in the best-paid occupations. The disproportion is even greater in the case of Hispanics, who are about 20 percent of the age group, but no more than 8 percent of the incumbents of good jobs. The absence of a substantial dent in white privilege can be seen as an expected feature of a non-zero-sum mobility situation, which entails little or no change to the perceptions whites have of the opportunities open to them and to their children.

What sorts of changes are to be anticipated by, say, 2020? If we assume that the top jobs of 2020 will be roughly those of 2000, overall change in their composition is programmed by the succession of cohorts, which will lead to the massive disappearance from the labor market of the job incumbents who were aged 45–64 in 2000. These are the cohorts most dominated by native-born non-Hispanic whites, and their places will be taken by the cohorts aged 25–44 in 2000, where the presence of nonwhites and Hispanics has increased markedly.

It is impossible to say for certain what the composition of younger cohorts will look like in 2020, in part because the answer depends on future immigration. But two observations point to further increases in the proportions of minorities in top jobs. One is the decline in the absolute number of native-born non-Hispanic whites available to take these jobs. This is particularly noticeable in the birth cohort that in 2020 will take the place of the 35–44-year-olds of 2000, as shown in table 19.3. The native-born white incumbents of top jobs in that age group were recruited from a population that, despite the mortality by early middle age, still

Table 19.3 U.S.-born, non-Hispanic white birth cohorts, counted in 2000 and projected for 2020

Age groups	2000 (1000s)	2020 (1000s)	% change
55–64	17,975.0	27,974.3	+55.6
45–54	26,947.2	22,991.4	−14.7
35–44	30,560.7	23,084.8	−24.5
25–34	23,910.4	24,464.3	+2.3
15–24	23,579.2	–	–
5–14	24,784.0	–	–

Note: Projections have been made by applying survival rates from the United States Life Tables for whites (NCHS, 2002).
National Center for Health Statistics (NCHS). (2002). United States Life Tables, 2000, NVSR Volume 51, Number 3: http://www.cdc.gov/nchs/products/pubs/pubd/nvsr/51/51_03.htm.

numbered nearly 31 million individuals. However, as of 2000, the 15–24-year-olds who will replace them comprised only 24 million native-born non-Hispanic whites, and mortality is likely to winnow this group somewhat between 2000 and 2020. By comparison, there is a rough stability of native-born whites between the 25–34-year-olds of 2000 and the cohort that forms their replacements, aged 5–14, in the census.

However, even population stability would not guarantee the white share of these jobs, whose number is likely to increase. Between 2000 and 2020, Census Bureau population projections suggest a total population increase on the order of 20 percent. Very likely, the occupations that we have designated as the top quartile and decile will expand in rough correspondence to the population, and the recruitment to them will have to expand accordingly. That will probably mean that the share of whites in these jobs will decline. The falloff is likely to be especially sharp among the 35–44-year-olds of 2020: if the recruitment of native-born whites to these jobs remains at the same proportion relative to the population base as it was in 2000, then there will only be enough of them to fill about 60 percent of the available positions (Alba 2009).

A drop this great seems an unlikely outcome, and in any event one cannot predict the future changes in top jobs with any precision. Perhaps, the recruitment of native-born whites to these jobs will involve a larger proportion of that population in the future than in the past. Perhaps the decline in the availability of qualified non-Hispanic whites will be made up by greater immigration. We don't know, but forecasting some degree of continuing decline in the non-Hispanic white share of the best-paid occupations seems a safe bet, considering the changes of the recent past and foreseeable demographic shifts.

Some Contingencies

Much has been made of the so-called hourglass economy, and there has been considerable attention to the expansion of the bad-job portion of the economy (Bernhardt et al. 2001). However, despite this pessimism, there are built-in demographic changes that, in combination with the institutional changes inaugurated by the civil rights era, are likely to bring a much larger number of nonwhites and Hispanics into the worlds of middle-class and upper-middle-class whites on a basis of rough parity. Should they occur, these changes would not mean an end to racial and ethnic inequalities. When groups are compared in the aggregate, they will continue to show large average disparities. But behind them, there is likely to be some reshuffling, as the overlaps between the overall distributions of white and minority status increase. This is another way of saying that a growing number of minorities will probably interact on a regular basis and as equals with whites as well as others whose origins are different from their own, and most whites will likely find themselves increasingly confronted with inescapable diversity. Such an interpenetration of social worlds is anticipated by assimilation theory.

The next few decades seem to offer then an extraordinary opportunity for minority mobility and for a reshuffling of the major racial/ethnic boundaries of U.S. society, which David Hollinger (1995) has described with the phrase "racial/ethnic pentagon." Yet, other than a more diverse mainstream than exists today, it is impossible to be very precise about the nature of the changes that may occur. This is in part because they will not be dictated by demographic and socioeconomic structures, which are to a great degree predictable, but be forged by human agents. Thus, among the questions that remain to be answered are the following:

1. To what extent will socioeconomically mobile minorities be able to realize broader social gains from their entry into higher-status occupational spheres? Assimilation in its broadest sense depends on the ability to integrate into mainstream social settings—to mix with whites and others of the same socioeconomic strata and to provide a favorable starting position for one's children. An enormous stream of research about African Americans shows that, because of discrimination and institutional racism, they have generally been unable to realize these gains—for instance, they have usually been confined to largely black residential areas regardless of their economic attainments (e.g., Massey and Denton 1993). What research there is suggests so far that Asians and light-skinned Hispanics are not as constrained in residential choice as blacks have been (Alba, Logan and Stults 2000, South, Crowder and Chavez 2005).

Intermarriage is another way of gaining insight into changes in social distance that may come about with rising socioeconomic status. Indicative is not simply the overall rate of intermarriage, but the gradient in that rate associated with a measure of socioeconomic position such as educational attainment. Here the picture is varied. Asian and Hispanic intermarriage with whites is very common among the highly educated; for the U.S.-born members of these groups, the rates hover in the 40–60 percent range depending on the precise category (Qian and Lichter 2007). By contrast, highly educated blacks do not improve their chances of intermarriage with whites above the low frequencies that obtain for the group as a whole.

2. To what extent will the decline in the number of whites in birth cohorts coming of age during the next two decades lead to enhanced opportunities for working-class whites rather than for minorities? An alternative scenario to a major racial/ethnic reordering in the near future is that white privilege will continue to exert a powerful hold on channels of mobility and that whites of lower socioeconomic origins will rise to occupy the positions that will open up. To be sure, some degree of racial/ethnic change is already visible in the shifts across cohorts within the highest occupational tiers, but future change will be constrained to the extent that whites can exploit new opportunities.

3. To what extent will white Americans be willing to invest in the education of minority and immigrant-origin youth to enable them to take advantage of opportunities for mobility? The public school system in the United States, especially in the predominantly minority areas, is increasingly in disarray, troubled by segregation and inadequacies of funding. Poor educational facilities will limit the ability of youth from low-income families to rise far above their original status.

4. To what extent will American society resort to the recruitment of immigrants trained partly or wholly in their home countries to fill jobs requiring high levels of qualification? Relying on immigrants is an alternative to the training of minority and immigrant-origin youth to replace the declining numbers of whites, and a cheap one at that, since much of the cost of educating highly qualified workers is borne by other countries. In recent years, the United States has expanded its intake of highly trained foreigners through the H-1B visa program, for example, which remains a tool that national policy makers can draw on. Our data show that immigrants are taking a disproportionate number of top jobs, indicating the viability of this alternative.

5. Will future changes affect African Americans and immigrant-origin groups equally? One of the profoundly rooted patterns in U.S. history is

the preference for immigrants over native minorities, especially African Americans (Waldinger 1996, Waldinger and Lichter 2003). This pattern operated during the twentieth century in the ability of once-despised southern and eastern European groups, such as the southern Italians, to distance themselves from black Americans and rise into the white American mainstream. This historical process is of course the subject of the whiteness literature, which has emphasized the devices employed by the immigrant ethnics to separate themselves from blacks and to gain acceptance as whites. There is a substantial risk of this pattern repeating itself in the contemporary era. This risk is visible in two ways: the continuing preference of white Americans for immigrants, who are seen as different from blacks in the degree to which they work hard to improve their lives and to provide opportunities for their children (Gans 1999, Waldinger and Lichter 2003), and the emerging tensions between immigrants and black Americans. A study of new Latin American immigrants in the smaller cities of upstate New York has found that the immigrants view African Americans as very hostile, while white Americans are perceived as welcoming or, at worst, neutral (Villarrubia, Denton and Alba 2007). Hence, the immigrants, who often live side by side with African Americans in very poor neighborhoods, are motivated to separate themselves from native minorities as soon as they can. This could solidify the African American/non-African American distinction as the key fault line in U.S. society.

Conclusion

The opportunity to alter the racial and ethnic boundaries of American society through increasing diversity at its middle and upper levels (achieved by the mobility into these tiers of the second generation and of native minorities) is, in the end, just that: an opportunity, not a sure thing. To realize it, the United States will need to invest more in the education of the young people from immigrant and minority backgrounds and to keep free the channels of upward mobility available to them. But the chance to realize significant racial and ethnic change does not come around very often. Promoting policies that will help to bring about such change seems a worthy goal for those who possess knowledge about the processes involved.

References

Abraham, Salim (2006). "Easing Foreign-Born Children into City School." Columbia News Service. Retrieved September 28 from http://jscms.jrn.columbia.edu/cns/2006-02-28
Adams, Henry (1918). *The Education of Henry Adams.* Boston: Houghton, Mifflin.
Addams, Jane (1910). *Twenty Years at Hull-House: With Autobiographical Notes.* New York: Macmillan.
Aguilera, Michael B., and Douglas S. Massey (2003). "Social Capital and the Wages of Mexican Migrants: New Hypotheses and Tests." *Social Forces* 82: 671–701.
Alba, Richard (1985). *Italian Americans: Into the Twilight of Ethnicity.* Englewood Cliffs: Prentice-Hall.
Alba, Richard (1995). "Assimilation's Quiet Tide." *The Public Interest* 119 (Spring): 1–18.
Alba, Richard (2009). *Blurring the Color Line: The New Chance for a More Integrated America.* Cambridge: Harvard University Press.
Alba, Richard, John Logan, and Brian Stults (2000). "The Changing Neighborhood Contexts of the Immigrant Metropolis." *Social Forces* 79 (December): 587–621.
Alba, Richard and Victor Nee (2003). *Remaking the American Mainstream: Assimilation and the New Immigration.* Cambridge: Harvard University Press.
Aleinikoff, T. Alexander (2001). "Policing Boundaries: Migration, Citizenship, and the State," pp. 267–291. In G. Gerstle and J. Mollenkopf (eds), *E Pluribus Unum? Contemporary and Historical Perspectives on Immigrant Political Incorporation.* New York: Russell Sage Foundation.
Aleinikoff, T. Alexander (2003). "Between National and Postnational: Membership in the United States," pp. 110–129. In C. Joppke and E. Morawska (eds.), *Toward Assimilation and Citizenship: Immigrants in Liberal Nation-States.* New York: Palgrave Macmillan.
Aluko, Remi and Diana Sherblom (1997). "Passing Culture on to the Next Generation: African Immigrant Language and Culture Schools in Washington, D.C.," *Festival of American Folklife Program Book.* Washington, D.C.: African Immigrant Folklife Project.
Arax, Mark (1987). "Refugees Called Victims and Perpetrators of Fraud." *Los Angeles Times*, Part I (February 9): 1, 3.
Arun Peter Lobo, "Unintended Consequences: Liberalized U.S. Immigration Law and the African Brain Drain," pp. 189–208. In Konadu-Agyemang, Kwado, Baffour Takyi and John Arthur (eds.), *The New African Diaspora in*

North America: Trends, Community Building, and Adaptation. Lanham, MD.: Lexington Books.

Ash, M. G., and A. Söllner (eds.) (1996). *Forced Migration and Scientific Change: Emigré German-Speaking Scientists and Scholars After 1933.* Cambridge, UK: Cambridge University Press.

Astone, N. M., and S. S. McLanahan (1991). "Family Structure, Parental Practices and High School Completion." *American Sociological Review* 56 (3): 309–320.

Attewell, Paul and David Lavin (2007). *Passing the Torch: Does Higher Education Pay Off Across the Generations.* New York: The Russell Sage Foundation.

Azikiwe, Nnamdi (1971). *My Odyssey.* London: C. Hurst and Co.

Bailey, Benjamin (2001). "Dominican-American Ethnic/Racial Identities and United States Social Categories," *International Migration Review* 35 (3): 677–708.

Bailyn, Bernard (1982). "The Challenge of Modern Historiography." *American Historical Review* 87: 1–24.

Bankston, C., and M. Zhou (2002). "Being Well vs. Doing Well: Self Esteem and School Performance Among Immigrant and Nonimmigrant Racial and Ethnic Group." *International Migration Review* 36 (2): 389–415.

Barkan, E. R. and N. Khokhlov (1980). "Socioeconomic Data as Indexes of Naturalization Patterns in the United-States—a Theory Revisited." *Ethnicity* 7: 159–190.

Barrett, James and David Roediger (1997). "Inbetween Peoples: Race, Nationality, and the 'New Immigrant' Working Class." *Journal of American Ethnic History* 16: 3–44.

Barton, Josef J. (1975). *Peasants and Strangers: Italians, Rumanians, and Slovaks in an American City, 1890–1950.* Cambridge, MA: Harvard University Press.

Basch, Linda, Nina Glick Schiller and Cristina Szanton Blanc (1994). *Nations Unbound: Transnational Projects, Postcolonial Predicaments, and Deterritorialized Nation States.* New York: Gordon and Breach.

Bates, Laura, Diane Baird, Deborah J. Johnson, Robert E. Less, T. Luster and C. Rehagen (2005). "Sudanese Refugee Youth in Foster Care: The 'Lost Boys' in America." *Child Welfare* 84 (September/October): 631–648.

Bauböck, Rainer (1994). *Transnational Citizenship: Membership and Rights in International Migration.* Aldershot: Edward Elgar.

Bean, Frank D., Susan K. Brown, and Rubén G. Rumbaut (2006). "Mexican Immigrant Political and Economic Incorporation." *Perspectives on Politics* 4: 309–313.

Bean, Frank D. and B. Lindsay Lowell (2004). "NAFTA and Mexican Migration to the United States," pp. 263–284. In Sidney Weintraub (ed.), *NAFTA'S Impact on North America: The First Decade.* Washington, D.C.: Center for Strategic and international Studies.

Bean, Frank D., Jeffrey S. Passel, and Barry Edmonston (1990). *Undocumented Migration to the United States: IRCA and the Experience of the 1980s.* Washington, D.C.: The Urban Institute Press.

Bean, Frank and Gillian Stevens (2003). *America's Newcomers and the Dynamics of Diversity.* New York: Russell Sage Foundation.

Bean, Frank D., Georges Vernez, and Charles B. Keely (1989). *Opening and Closing the Doors: Evaluating Immigration Reform and Control*. Santa Monica, CA., and Washington, D.C.: The RAND Corporation and the Urban Institute.
Beijbom, U. (1971). *Swedes in Chicago: A Demographic and Social Study of the 1846–1880 Immigration*. Chicago: Chicago Historical Society.
Bergad, Laird (n.d.). "Comparative Income Distribution Patters Among Hispanic Households in New York City," *Census 2000*. Accessed at: http://web.gc.cuny.edu/lastudies. April 1, 2010.
Bernard, W. S. (1936). "Cultural Determinants of Naturalization." *American Sociological Review* 1: 943–953.
Bernhardt, Annette, Martina Morris, Mark Handcock, and Marc Scott (2001). *Divergent Paths: Economic Mobility in the New American Labor Market*. New York: Russell Sage Foundation.
Berrol, Selma (1974). "Turning Little Aliens into Little Citizens: Italians and Jews in New York City Public Schools, 1900–1914." *Proceedings of the Seventh Annual Conference of the American Italian Historical Association*. Towson State College, Md.
Berthoff, Rowland (1971). *An Unsettled People: Social Order and Disorder in American History*. New York: Harper and Row.
Berthoff, Rowland (1997). *Republic of the Dispossessed: The Exceptional Old-European Consensus in America*. Columbia: University of Missouri Press.
Bhabha, J. and S. Schmidt (2006). *Seeking Asylum Alone: Unaccompanied and Separated Children and Refugee Protection in the U.S.* Cambridge, MA: University Committee on Human Rights Studies, Harvard University.
Blau, Peter, and Otis Duncan (1967). *The American Occupational Structure*. New York: Academic Press.
Bloemraad, Irene (2004). "Who Claims Dual Citizenship? The Limits of Postnationalism, the Possibilities of Transnationalism, and the Persistence of Traditional Citizenship." *International Migration Review* 38: 5–42.
Bloemraad, Irene (2005). "The Limits of Tocqueville: How Government Facilitates Organisational Capacity in Newcomer Communities." *Journal of Ethnic and Migration Studies* 31 (5): 865–887.
Bloemraad, Irene (2006). *Becoming a Citizen: Incorporating Immigrations and Refugees in the United States and Canada*. Berkeley: University of California Press.
Bodnar, John (1985). *The Transplanted: A History of Immigrants in Urban America*. Bloomington: Indiana University Press.
Boris, Elizabeth and Eugene Steuerle (1999). *Nonprofits and Government: Collaboration and Conflict*. Washington, D.C.: Urban Institute Press.
Borjas, George J. (2001). *Heaven's Door: Immigration Policy and the American Economy*. Princeton, N.J.: Princeton University Press.
Boston, C. (2003). *High School Report Cards* ERIC Digest.
Bourdieu, P. (1983). "Ökonomisches Kapital, kulturelles Kapital, soziales Kapital," pp. 183–198. In R. Kreckel (ed.), *Soziale Ungleichheiten*. Göttingen: Schwartz.

Bowen, William and Derek Bok (1998). *The Shape of the River: Long-Term Consequences of Considering Race in College and University Admissions.* Princeton: Princeton University Press.

Boyce Rodgers, K., and H. A. Rose (2001). "Personal, Family, and School Factors related to Adolescent Academic Performance: A Comparison By Family Structure." *Marriage and Family Review* 33 (4): 47–61.

Bozorgmehr, Mehdi (2007). "Iran," pp. 469–478. In Mary Waters and Reed Ueda with Helen B. Marrow (eds), *The New Americans: A Guide to Immigration Since 1965.* Cambridge, Harvard University Press.

Brandon, P. (1991). "Gender Differences in Young Asian Americans' Educational Attainment." *Sex Roles* 25 (1/2): 45–61.

Branscomb, L., G. Holton, and G. Sonnert (2002). "Science for Society," pp. 397–433. In A. H. Teich, S. D. Nelson & S. J. Lita (eds), *AAAS Science and Technology Policy Yearbook 2002.* Washington, D.C.: American Association for the Advancement of Science.

Brimelow, Peter (1995). *Alien Nation: Common Sense About America's Immigration Disaster.* New York: Random House.

Breton, Raymond (1964). "Institutional Completeness of Ethnic Communities and the Personal Relations of Immigrants." *American Journal of Sociology* 70 (2): 193–205.

Brookhart, S. M. (1994). "Teachers' Grading: Practice and Theory." *Applied Measurement in Education* 7 (4): 279–301.

Bryk, A., and B. Schneider (2003). "Trust in Schools: A Core Resource for School Reform." *Educational Leadership* 60 (6): 40–44.

Burr, Clinton Stoddard (1922). *America's Race Heritage.* New York: National Historical Society.

Burt, Ronald S. (2004). "Structural Holes and Good Ideas." *American Journal of Sociology* 110: 349–399.

Camarota, Steven A. (2001). "Immigration From Mexico: Assessing the Impact on the United States." Washington, D.C.: Center for Immigration Studies. Accessed at: http://www.cis.org/articles/2001/Mexico/toc.html. April 30, 2010.

Camarota, Steven A. (2002). *Immigrants in the United States—2002 A Snapshot of America's Foreign-Born Population.* Staten Island, NY: Center for Migration Studies. Accessed at: http://www.cis.org/articles/2002/back1302.html. September 12, 2006.

Caplan, Nathan, Marcella H. Choy and John K. Whitmore (1992). "Indochinese Refugee Families and Academic Achievement." *Scientific American* (February): 36–42.

Carens, Joseph H. (1987). "Aliens and Citizens: The Case for Open Borders." *The Review of Politics* 49: 251–273.

Carhill, A., M. Páez, and C. Suárez-Orozco (2008). "Explaining English Language Proficiency Among Adolescent Immigrant Students." *American Educational Research Journal* 45 (4): 1155–1179.

Carter, Prudence (2005). *Keepin' It Real: School Success beyond Black and White.* New York: Oxford University Press.

Castles, Stephen and Mark J. Miller (1993). *The Age of Migration: International Population Movements in the Modern World*. New York: Guilford.

Cauce, A. M., R. D. FelnerD, and J. Primavera (1982). "Social Support in High-risk Adolescents: Structural Components and Adaptive Impact." *American Journal of Community Psychology* 10 (4): 417–428.

Chung, Angie Y. (2005). " 'Politics Without the Politics': The Evolving Political Cultures of Ethnic Non-profits in Koreatown, Los Angeles." *Journal of Ethnic and Migration Studies* 31 (5): 911–929.

Cohen, Miriam (1992). *Workshop to Office: Two Generations of Italian Women in New York, 1900–1950*. Ithaca: Cornell University Press.

Collier, V. P. (1992). "A Synthesis of Studies Examining Long-Term Language-Minority Student Data on Academic Achievement." *Bilingual Research Journal* 16 (1 & 2): 187–212.

Cordero-Guzman, Hector (2005). "Community Based Organisations and Migration in New York City." *Journal of Ethnic and Migration Studies* 31 (5): 889–909.

Coser, L. (1984). *Refugee Scholars in America: Their Impact and Their Experiences*. New Haven, CT: Yale University Press.

Counts, George S. (1922). *The Selective Character of American Secondary Education*. Chicago: University of Chicago Press.

Covello, Leonard (1967). *The Social Background of the Italo-American School Child*. Leiden: Brill.

Crul, M., and J. Doomernik (2003). "The Turkish and Moroccan Second Generation in the Netherlands: Divergent Trends Between and Polarization Within the Two Groups." *Internation Migration Review* 37 (4): 1039–1064.

Cummins, J. (1991). "Language Development and Academic Learning," pp. 161–175. In L. M. Malavé & G. Duquette (eds), *Language, Culture, & Cognition*. Clevedon, England: Multilingual Matters.

DeCapua, A., W. Smathers and L. F. Tang (2007). "Schooling Interrupted." *Educational Leadership* 64 (6): 40.

Delgado, Hector (1993). *New Immigrants, Old Unions: Organizing Undocumented Workers in Los Angeles*. Philadelphia: Temple University Press.

Deschenes, Sarah, Milbrey McLaughlin and Jennifer O'Donahue (2006). "Nonprofit Community Organizations in Poor Urban Settings: Bridging Institutional Gaps for Youth." In Walter Powell and Richard Steinberg (eds), *The Nonprofit Sector: A Research Handbook*. New Haven: Yale University Press.

DeSipio, Louis, Frank D. Bean, and Rubén G. Rumbaut (2005). "Immigration Status and Naturalization Across Generations: The Consequences of Parental Unauthorized Migration or Naturalization on the Civic and Political Behaviors of 1.5 and 2nd Generation Young Adults in Los Angeles." Paper presented at the Annual Meetings of the American Political Science Association, Washington, D. C.: September.

Dinnerstein, Leonard (1994). *Anti-Semitism in America*. New York: Oxford University Press.

Donnelly, Ignatius (1890). *Caesar's Column*. Chicago: F. J. Schulte.

Dozier, Sandra Bygrave (2001). "Undocumented and Documented International Students: A Comparative Study of their Academic Performance." *Community College Review*. Accessed at: findarticles.com.

Du Bois, W. E. B. (1903). *The Souls of Black Folk Essays and Sketches*. Chicago: A. C. McClurg.

Dubow, E. F. (1991). "A Two-Year Longitudinal Study of Stressful Life Events, Social Support, and Social Problem-Solving Skill: Contributions to Children's Behavioral and Academic Adjustment." *Child Development* 65: 583–599.

Dunn, Ashley (1995). "Cram Schools: Immigrants' Tools for Success." *New York Time*, January 28.

Eccles, J. S., A. Wigfield and U. Schiefele(1998). "Motivation to Succeed," 5th ed., Vol. 3, pp. 1017–1095. In W. Damon & N. Eisenberg (eds.), *Handbook of Child Psychology*. New York: John Wiley and Sons, Inc.

Elliott, D. S., B. A. Hamburg, and K. R. Williams (eds.) (1998). *Violence in American Schools*. New York, NY: Cambridge University Press.

Ellis, B. H. (2007). "Trauma and Mental Health in Child and Adolescent Refugees," *Center for Behavioral Health*. Retrieved October 1, 2007 from http://www.nctsnet.org/nctsn_assets/pdfs/Child_Refugee_Mental_Health_Ellis8-07.pdf

Emmons, S. and D. Reyes (1989). "Gangs, Crime Top Fear of Vietnamese in Orange County." *Los Angeles Times*, February 5.

Espiritu, Yen Le (1992). *Asian American Pan-Ethnicity: Bridging Institutions and Identities*. Philadelphia, Temple University Press.

Espiritu, Yen Le (2002). "The Intersection of Race, Ethnicity and Class: The Multiple Identities of Second Generation Filipinos, pp. 19–52." In Pyong Gap Min (ed.), *Second Generation: Ethnic Identity Among Asian Americans*. New York: Altamira Press.

Featherman, David, and Robert Hauser (1978). *Opportunity and Change*. New York: Academic Press.

Ferber, Thaddeus, Elizabeth Gaines, and Christi Goodman (October 2005). *Positive Youth Development: State Strategies. Strengthening State Policy: Research and Policy Report*. Washington, D.C.: National Conference of State Legislatures.

Feldblum, Miriam (2000). "Managing Membership: New Trends in Citizenship and Nationality Policy," pp. 475–499. In T. A. Aleinikoff and D. Klusmeyer (eds.), *From Migrants to Citizens: Membership in a Changing World*. Washington, D.C.: Carnegie Endowment for International Peace.

Fermi, L. (1971). *Illustrious Immigrants: The Intellectual Migration from Europe, 1930–41*. Revised edition. Chicago: University of Chicago Press.

Fleming, Donald (1963). "Social Darwinism," pp. 123–130. In Arthur M. Schlesinger, Jr. and Morton White (eds.), *Paths of American Thought*. Boston: Houghton Mifflin.

Fleming, D., and B. Bailyn (eds.) (1969). *The Intellectual Migration: Europe and America, 1930–1960*. Cambridge, MA: Belknap Press.

Foner, Nancy (2000). *From Ellis Island to JFK: New York's Two Great Waves of Immigration*. New Haven and New York: Yale University Press and Russell Sage Foundation.

Foner, Nancy (2005). *In a New Land: A Comparative View of Immigration.* New York: New York University Press.

Foner, Nancy and Richard Alba (2006). "The Second Generation from the Last Great Wave of Immigration: Setting the Record Straight." *Migration Information Source.* http://www.migrationinformation.org/Feature/display.cfm?id=439.

Foner, Nancy and George Fredrickson (2004). "Immigration, Race, and Ethnicity in the United States: Social Constructions and Social Relations." In Nancy Foner and George Fredrickson (eds.), *Not Just Black and White: Historical and Contemporary Perspectives on Immigration, Race, and Ethnicity in the United States.* New York: Russell Sage Foundation.

Foner, Nancy and Philip Kasinitz (2007). "The Second Generation." In Mary C. Waters and Reed Ueda (eds.), *The New Americans.* Cambridge: Harvard University Press.

Franco, Adrian (2007). Personal Communication. February 12, 2007.

Fredricks, J. A., P. C. Blumenfeld, and A.H. Paris (2004). "School Engagement: Potential of the Concept, State of the Evidence." *Review of Educational Research* 74 (1): 54–109.

Fuchs, Lawrence (1982). *The American Kaleidoscope: Race, Ethnicity and the Civic Culture.* Hanover, New Hampshire: University Press of New England.

Fuligini, A. (1997). "The Academic Achievement of Adolescents from Immigrant Families: The Roles of Family Background, Attitudes, and Behavior." *Child Development* 69 (2): 351–363.

Fuligini, A., and S. Pederson (2002). "Family Obligation and the Transition to Young Adulthood." *Developmental Psychology* 38 (5): 856–868.

Furstenberg, Frank, Thomas D. Cook, Jaquelynne Eccles, Glen Elder Jr., and Arnold Sameroff (1999). *Managing to Make It: Urban Families and Adolescent Success.* Chicago: University of Chicago Press.

Gans, Herbert (1992). "Second Generation Decline: Scenarios for the Economic and Ethnic Futures of Post-1965 American Immigrants." *Ethnic and Racial Studies* 15 (April): 173–192.

Gans, Herbert (1999). "The Possibility of a New Racial Hierarchy in the Twenty-first Century United States." In Michele Lamont (ed.), *The Cultural Territories of Race: Black and White Boundaries.* Chicago: University of Chicago Press.

Garbarino, J., and N. Dubrow (1997). *Children in Danger: Coping with the Community Consequences of Violence.* San Francisco: Jossey-Bass.

García-Coll, C., and K. Magnuson (1997). "The Psychological Experience of Immigration: A Developmental Perspective," pp. 91–132. In A. Booth, A. C. Crouter & N. Landale (eds.), *Immigration and the Family.* Mahwah: Lawrence Erlbaum.

Gates Foundation (2006). *The 3Rs Solution.* Retrieved August 1, 2006, from http://www.gatesfoundation.org/Education/RelatedInfo/3Rs_Solution.htm

Geltman, Paul L. et al. (2003). *A Nationwide Study of the Functional and Behavioral Health of Sudanese Unaccompanied Refugee Minors Resettled in the United States.* Executive Summary. April 2003. Available on www.brycs.org.

George, Nelson (1998). *Hip Hop America.* New York, NY: Penguin: 121–127.

Gilbertson, Greta A. and Audrey Singer (2003). "The Emergence of Protective Citizenship in the USA: Naturalization Among Dominican Immigrants in the Post-1996 Welfare Reform Era." *Ethnic and Racial Studies* 26: 25–51.
Goldmark, Pauline (1914). *West-Side Studies*, Vol. I. New York: Russell Sage Foundation.
Glazer, Nathan (1997). *We are All Multiculturalists Now*. Cambridge: Harvard University Press.
Glazer, Nathan, and Daniel P. Moynihan (1963). *Beyond the Melting Pot: The Negroes, Puerto Ricans, Jews, Italians, and Irish of New York City*. Cambridge, MA: M.I.T. Press and Harvard University Press.
Glazer, Nathan and Daniel Patrick Moynihan (1970). *Beyond the Melting Pot*. 2nd ed. Cambridge: MIT Press.
Glick Schiller, Nina, Linda Basch, and Cristina Blanc-Szanton (1992). *Towards a Transnational Perspective on Migration: Race, Class, Ethnicity, and Nationalism Reconsidered*. New York, N.Y.: New York Academy of Sciences.
Glick Schiller, Nina (1999). "Transmigrants and Nation-States: Something Old and Something New in the U.S. Immigrant Experience," pp. 94–119. In *The Handbook of International Migration: The American Experience* (eds.), Charles Hirschman, Philip Kasinitz, and Josh DeWind. New York: Russell Sage.
Gold, Steven (1992). *Refugee Communities: A Comparative Field Study*. Newbury Park, CA: Sage.
Gold, Steven (1994). "Chinese-Vietnamese Entrepreneurs in California," pp. 196–226. In Paul Ong, Edna Bonacich and Lucy Cheng (eds.), *The New Asian Immigration in Los Angeles and Global Restructuring*. Philadelphia: Temple University Press.
Gold, Steven (1995). *From The Workers' State to The Golden State: Jews from the Former Soviet Union in California*. Boston: Allyn and Bacon.
Gold, Steven (1999). "Southeast Asians," pp. 505–519. In Elliot Barkan (ed.), *Our Multicultural Heritage: A Guide to America's Principal Ethnic Groups*. Santa Barbara, CA: Greenwood Press.
Gold, Steven and Mehdi Bozorgmehr (2007). "Middle East and North Africa," pp. 518–533. Mary Waters and Reed Ueda with Helen B. Marrow (eds.), *The New Americans: A Guide to Immigration Since 1965*. Cambridge: Harvard University Press.
González Baker, Susan (1990). *The Cautious Welcome*. Santa Monica, CA and Washington, D.C.: The RAND Corporation and The Urban Institute.
González Baker, Susan (1997). "The 'Amnesty' Aftermath: Current Policy Issues Stemming from the Legalization Programs of the 1986 Immigration Reform and Control Act." *International Migration Review* 31: 5–27.
Gonzalez, E. (2006, June). *Response to Recommendation to Limit USCIS Adjudication of Asylum Applications to Those Submitted by Individuals in Valid Non-Immigrant Status*. Paper presented electronically. http://www.dhs.gov/xlibrary/assets/CISOmbudsman_RR_24_Asylum_Status_USCIS_Response-06-20-06.pdf
González-López, Gloria (2003). "*De madres a hijas*: Gendered Lessons on Virginity across Generations of Mexican Immigrant Women," pp. 217–240. In Pierrette

Hondagneu-Sotelo (ed.), *Gender and U.S. Immigration: Contemporary Trends*. Berkeley: University of California Press.

Goodenow, C. (1993). "Classroom Belonging Among Early Adolescent Students: Relationships to Motivation and Achievement." *Journal of Early Adolescence* 13 (1): 21–43.

Gordon, April (1998). "The New Diaspora—African Immigration to the United States." *Journal of Third World Studies* 15 (1): 79–103.

Gordon, Milton (1964). *Assimilation in American Life: The Role of Religion and National Origins*. New York: Oxford University Press.

Gorelick, Sherry (1981). *City College and the Jewish Poor: Education in New York, 1880–1924*. New Brunswick, Rutgers University Press.

Gowan, Mary, and Melanie Treviño (1998). "An Examination of Gender Differences in Mexican-American Attitudes Toward Family and Career Roles." *Sex Roles* 38: 1079–1093.

Gozdziak, Elzbieta M. and Micah Bump (2007). *Victims No Longer: Research on Child Survivors of Trafficking for Sexual and Labor Exploitation*. Washington, D.C.: National Institute of Justice.

Granovetter, Mark (1973). "The Strength of Weak Ties." *American Journal of Sociology* 78: 360–380.

Grebler, Leo, Joan W. Moore, and Ralph C. Guzman (1970). *The Mexican-American People: The Nation's Second Largest Minority*. New York: Free Press.

Greenwood, C. R., B. T. Horton, and C. A. Utley (2002). "Academic Engagement: Current Perspectives in Research and Practice." *School Psychology Review* 31 (3): 1–31.

Guarnizo, Luis, Alejandro Portes, and William J. Haller (2003). "Assimilation and Transnationalism: Determinants of Transnational Political Action among Contemporary Migrants." *American Journal of Sociology* 108 (6): 1211–1248.

Gutman, Herbert G. (1973). "Work, Culture, and Society in Industrializing America, 1815–1919," *American Historical Review* 78 (June): 531–588.

Hagan, Jacqueline and Susan Gonzalez Baker (1993). "Implementing the US Legalization Program: the Influence of Immigrant Communities and Local Agencies on Immigration Policy Reform." *International Migration Review* 27 (3): 513–535.

Haines, David (2007). "Refugees," pp. 56–69. In Mary Waters and Reed Ueda with Helen B. Marrow (eds.), *The New Americans: A Guide to Immigration Since 1965*. Cambridge, Harvard University Press.

Halter, Marilyn (1993). *Between Race and Ethnicity: Cape Verdean American Immigrants, 1865–1960*. Urbana-Champaign, IL: University of Illinois Press.

Handlin, Oscar (1959). *Immigration as a Factor in American Life*. Englewood Cliffs, N. J.: Prentice Hall.

Hansen, Marcus Lee (1938/1990). "The Problem of the Third Generation Immigrant." In Peter Kivisto and Dag Blanck (eds.), *American Immigrants and Their Generations: Studies and Commentaries on the Hansen Thesis After Fifty Years*. Urbana: University of Illinois Press (Originally published in Augustana Historical Society Publications.).

Hanson, Victor Davis (2003). *Mexifornia: A State of Becoming*. San Francisco: Encounter Books.

Harzig, Christiane (2006). "Domestics of the World (Unite?): Labor Migration Systems and Personal Trajectories of Household Workers in Historical and Global Perspective." *Journal of American Ethnic History* 25: 48–73.
Heckman, J. J. (1992). "Randomization and Social Policy Evaluation," pp. 201–230. In C. F. Manski and I. Garfinkel (eds.), *Evaluating Welfare and Training Programs*. Cambridge: Harvard University Press.
Heckman, J. J. (1996). "Randomization as an Instrumental Variable." *Review of Economic Statistics* 77: 336–341.
Heckman, J. J. (1997). "Instrumental Variables: A Study of Implicit Behavioral Assumptions Used in Making Program Evaluations." *Journal of Human Resources* 32: 441–462.
Heckman, J. J., J. Smith, and N. Clements (1997). "Making the Most Out of Programme Evaluations and Social Experiments: Accounting for Heterogeneity in Programme Impacts." *Review of Economic Studies* 64: 487–535.
Heilbut, A. (1983). *Exiled in Paradise: German Refugee Artists and Intellectuals in America, From the 1930s to the Present*. New York: Viking Press.
Hein, Jeremy (2006). *Ethnic Origins: The Adaptation of Cambodian and Hmong Refugees in Four American Cities*. New York: Russell Sage Foundation.
Higham, John (1955). *Strangers in the Land: Patterns of American Nativism, 1860–1925*. New Brunswick: Rutgers University Press.
Hirsch J., and E. Hirsch (1961). "Berufliche Eingliederung und wirtschaftliche Leistung der deutsch-jüdischen Einwanderung in die Vereinigten Staaten," pp. 41–70. In American Federation of Jews from Central Europe, *Twenty Years American Federation of Jews from Central Europe, Inc. 1940–1960*, New York: American Federation of Jews from Central Europe, Inc.
Hirschman, Charles (2005). "Immigration and the American Century." *Demography* 42: 595–620.
Hofstadter, Richard (1948). *The American Political Tradition*. New York: Alfred A. Knopf.
Hollinger, David (1995). *Postethnic America: Beyond Multiculturalism*. New York: Basic Books.
Holton, G., and G. Sonnert (1999). "A Vision of Jeffersonian Science." *Issues in Science and Technology* 16 (1): 61–65.
Hondagneu-Sotelo, Pierrette (1994). *Gendered Transitions: Mexican Experiences of Immigration*. Berkeley: University of California Press.
Hughes, H. S. (1975). *The Sea Change: The Migration of Social Thought, 1930–1965*. New York: Harper & Row.
Hulpia, H., and M. Valcke (2004). "The Use of Performance Indicators in a School Improvement Policy: The Theoretical and Empirical Context." *Evaluation and Research in Education* 18 (1–2): 102–119.
Hunter, Robert (1904). *Poverty: Social Conscience in the Progressive Era*. New York: Macmillan.
Huntington, Samuel P. (2004). *Who Are We? : The Challenges to America's Identity*. New York: Simon and Schuster.
Hutchison, William R. (2005). "Strong Objections: Another Best-Selling Author Complains about Plagiarism," *History News Newsletter* (December 5).

Ingrisch, D. (2004). *Der dis/kontinuierliche Status des Seins: Über vom Nationalsozialismus aus Österreich vertriebene (und verbliebene) intellektuelle Kulturen in lebensgeschichtlichen Kontexten*. Frankfurt am Main: Peter Lang Europäischer Verlag der Wissenschaften.

International Save the Children Alliance (2006). *Rewrite the Future: Education for Children in Conflict-affected Countries* (London: Cambridge House).

Iorizzo, Luciano J. and Salvatore Mondello (1980). *The Italian Americans*. Boston: G.K. Hall.

Itzigsohn, Jose and Silvia Giorguli Saucedo (2002). "Immigrant Incorporation and Sociocultural Transnationalism." *International Migration Review* 36 (3): 766–798.

Jackman, J. C., and C. M. Borden (1983). *The Muses Flee Hitler: Cultural Transfer and Adaption, 1930–1945*. Washington, D.C.: Smithsonian Institution Press.

Jacobson, David (1996). *Rights Across Borders: Immigration and the Decline of Citizenship*. Baltimore: Johns Hopkins University Press.

Jacobson, Matthew Frye (1998). *Whiteness of a Different Color: European Immigrants and the Alchemy of Race*. Cambridge: Harvard University Press.

Jasso, Guillermina, and Mark R. Rosenzweig (1986). "Family Reunification and the Immigration Multiplier: U.S. Immigration Law, Origin-Country Conditions, and the Reproduction of Immigrants." *Demography* 23: 294–311.

Jefferys, Kelly (2006). "Refugees and Asylees: 2005." *Office of Immigration Statistics Policy Directorate, Annual Flow Report*.

Jencks, C. (1972). *Inequality: A Reassessment of the Effect of Family and Schooling in America*. New York: Harper Colophon.

Jenkins, P. H. (1997). "School Delinquency and the School Social Bond." *Journal of Research in Crime & Delinquency* 34 (3): 337–367.

Johnson, Violet Showers (2006). *The Other Black Bostonians: West Indians in Boston, 1900–1950*. Bloomington: Indiana University Press.

Joyce, M. S. and W. A. Schambra (1996). "A New Civic Life, pp. 11–29." In M. Novak (ed.), *To Empower People: From State to Civil Society*. Washington D.C.: AEI Press.

Kao, G., and M. Tienda (1995). "Optimism and Achievement: The Educational Performance of Immigrant Youth." *Social Science Quarterly* 76 (1): 1–19.

Karabel, Jerome (2005). *The Chosen: The Hidden History of Admission and Exclusion at Harvard, Yale, and Princeton*. Boston: Houghton Mifflin.

Karatzias, A., K. G. Power, J. Flemming, and F. Lennan (2002). "The Role of Demographics, Personality Variables and School Stress on Predicting School Satisfaction/Dissatisfaction: Review of the Literature and Research Findings." *Educational Psychology* 22 (1): 33–50.

Kasinitz, Phillip (1992). *Caribbean New York: Black Immigrants and the Politics of Race*. Ithaca: Cornell University Press.

Kasinitz, Philip, John Mollenkopf and Mary C. Waters (2002). "Becoming American/ Becoming New Yorkers: The Experience of Assimilation in a Majority Minority City." *International Migration Review* 36 (4) (Winter): 1020–1036.

Kasinitz, Philip, John H. Mollenkopf, and Mary C. Waters (eds.) (2004). *Becoming New Yorkers: Ethnographies of the New Second Generation*. New York: Russell Sage.

Kasinitz, Philip, John H. Mollenkopf, Mary C. Waters and Jennifer Holdaway (2008). *Inheriting the City: The Children of Immigrants Come of Age*. Cambridge, MA: Harvard University Press and the Russell Sage Foundation.

Kasson, John F. (1978). *Amusing the Million: Coney Island at the Turn of the Century*. New York: Hill and Wang.

Katznelson, Ira (2005). *When Affirmative Action Was White: An Untold History of Racial Inequality in Twentieth-Century America*. New York & London: W.W. Norton.

Kenny-Benson, G. A., E. M. Pomerantz, A. M. Ryan, and H. Patrick (2006). "Sex Differences in Math Performance: The Role of Children's Approach to Schoolwork." *Developmental Psychology* 42 (1): 11–26.

Kent, D. (1953). *The Refugee Intellectual*. New York: Harper & Brothers.

Kessner, Thomas (1977). *The Golden Door: Italian and Jewish Immigrant Mobility in New York City, 1880–1915*. New York: Oxford University Press.

Kett, Joseph F. (1977). *Rites of Passage: Adolescence in America 1790 to the Present*. New York: Basic Books.

Kibria, Nazli (1993). *Family Tightrope: The Changing Lives of Vietnamese Americans*. Princeton, NJ: Princeton University Press.

Kibria, Nazli (2002). *Becoming Asian American: Second Generation Chinese and Korean American Identities*. Baltimore: Johns Hopkins University Press.

Klesmer, H. (1994). "Assessment & Teacher Perceptions of ESL Student Achievement." *English Quarterly* 26 (3): 8–11.

Lamont, Michele (1999). "Introduction: Beyond Taking Culture Seriously." In Michele Lamont (ed.), *The Cultural Territories of Race*. Chicago: University of Chicago Press.

Lavin, David and David Hyllegard (1996). *Changing the Odds: Open Addmissions and the Life Chances of the Disadvantaged*. New Haven: Yale University Press.

Layton, Lyndsey (2006). "More Refugees Are Settling in Mid-Size Cities, Study Finds." *Washington Post*, September 28: A03.

Lee, S. (1996). *Unraveling the "Model Minority" Stereotype: Listening to Asian American Youth*. New York: Teachers College Press.

Levitt, Peggy (2002). "The Ties that Change: Relations to the Ancestral Home Over the Life Cycle," pp. 123–144. In Levitt and Waters (eds.), *The Changing Face of Home: The Transnational Lives of the Second Generation*. New York: Russell Sage Foundation.

Levitt, Peggy, Josh DeWind, Steven Vertovec (2003). "International Perspectives on Transnational Migration: An Introduction." *International Migration Review* 37 (3): 565–575.

Levitt, Petty and Mary C. Waters (ed.) (2002). *The Changing Face of Home: The Transnational Lives of the Second Generation*. New York: Russell Sage Foundation.

Liang, Zai (1994). "Social Contact, Social Capital, and the Naturalization Process: Evidence from Six Immigrant Groups." *Social Science Research* 23: 407–437.

Lieberson, Stanley (1980). *A Piece of the Pie: Blacks and White Immigrants since 1880.* Berkeley: University of California Press, 1980.
Lieberson, Stanley and Mary Waters (1988). *From Many Strands: Ethnic and Racial Groups in Contemporary America.* New York: Russell Sage Foundation.
Logan, John, Brian Stults, and Reynolds Farley (2004). "Segregation of Minorities in the Metropolis: Two Decades of Change." *Demography* 41: 1–22.
London, Jack (1908). *The Iron Heel.* New York: Macmillan.
Lopez, David (2008). "Whither the Flock? The Catholic Church and the Success of Mexicans in America." In Richard Alba, Albert Raboteau and Josh DeWind (eds.), *Religion, Immigration and Religion in America.* New York: New York University Press.
Lopez, N. (2003). *Hopeful Girls, Troubled Boys: Race and Gender Disparity in Urban Education.* NY: Routledge.
Luster Tom, Deborah J. Johnson, and Laura Bates (2005). " 'Lost Boys' Finding Their Way: Challenges, Changes and Small Victories of Young Sudanese Refugees in the United States," Department of Family and Child Ecology, Michigan State University, East Lansing, MI.
Luster, T., L. Bates, and D. J. Johnson (2006). "Risk and Resilience," Vol. 2, pp. 27–52. In F. Villarruel and T. Luster (eds.), *Disorders in Adolescence.* In H. E. Fitzgerald, R. Zucker, & K. Feerark (Series eds.), *The Crisis in Youth Mental Health.* Praeger, Westport, CT.
Lynn, McBrien, J. (2006). "Educational Needs and Barriers for Refugee Students in the United States: A Review of the Literature." *Review of Educational Research* 75 (3): 333.
MacDonald, Heather (1995). "Why Koreans Succeed." *City Journal* 5: 12–29.
Marshall, Wayne (2005). "Hearing Hip-Hop's Jamaican Accent," *Newsletter of the Institute for Studies in American Music* XXXIV (2 spring): 8–9, 14–15.
Martin, David (2005). *US Refugee Program in Transition. Migration Information Source.* www.migrationinformation.org/Features/print.cfm?1D-305.
Massey, Douglas S. (1987). "Do Undocumented Migrants Earn Lower Wages than Legal Immigrants? New Evidence from Mexico." *International Migration Review* 2: 236–274.
Massey, Douglas (2002). *The River: The Social Origin of Freshmen at America's Selective Colleges and Universities.* Princeton: Princeton University Press.
Massey, Douglas, Camille Z. Charles, Garvey Lundy, and Mary Fischer (2003). *The Source of the River: The Social Origins of Freshmen at America's Selective Colleges and Universities.* Princeton: Princeton University Press.
Massey, Douglas and Nancy Denton (1993). *American Apartheid: Segregation and the Making of the Underclass.* Cambridge: Harvard University Press.
Massey, Douglas S., Jorge Durand, and Nolan J. Malone (2002). *Beyond Smoke and Mirrors: Mexican Immigration in an Era of Economic Integration.* New York: Russell Sage Foundation.
Massey, Douglas, Margarita Mooney, Kimberly C. Torres and Camille Z. Charles (2007). "Black Immigrants and Black Natives Attending Selective Colleges and Universities in the United States." *American Journal of Education* 113 (February): 243–271.

Matthews, Glenna (1987). *"Just a Housewife": The Rise and Fall of Domesticity in America*. New York: Oxford University Press.
Mauch, Christof and Joseph Salmons (eds.) (2003). *German-Jewish Identities in America*. Madison, WI: Max Kade Institute for German-American Studies.
McGill, L. D. (2005). *Constructing Black Selves: Caribbean American Narratives and the Second Generation*. New York: New York University Press.
McLaughlin, Milbrey, Merita Irby and Juliet Langman (2001). *Urban Sanctuaries: Neighborhood Organizations in the Lives and Futures of Inner City Youth*. San Francisco: Jossey-Bass.
McLaughlin, M. W., and J. E. Talbert (1993). *Contexts That Matter for Teaching and Learning*. Stanford, CA: Center for Research on the Context of Teaching at the Stanford University School of Education.
McNees, M., N. Siulc, O. Flores, and R. Smith (2003). *Mexican Immigrant Health in New York*. New York: Report to the United Hospital Foundation.
Mehan, H., I. Villanueva, L. Hubbard, and A. Lintz (1996). *Constructing School Success: The Consequences of Untracking Low Achieving Students*. New York: Cambridge University Press.
Midgley, C., H. Feldlaufer and J. S. Eccles (1989). "Student/Teacher Relations and Attitudes Toward Mathematics Before and After the Transition to Junior High School." *Child Development* 60: 981–992.
Milkman, Ruth (2006). *LA Story: Immigrant Workers and the Future of the U.S. Labor Movement*. New York: Russell Sage Foundation.
Min, Pyong Gap (1998). *Changes and Conflicts: Korean Immigrant Families in New York*. Boston: Allyn and Bacon.
Min, Pyong Gap (2001). "Koreans: An 'Institutionally Complete Community' in New York." In Nancy Foner (ed.), *New Immigrants in New York*. Rev. ed. New York: Columbia University Press.
Min, P. G. (2002). *The Second Generation: Ethnic Identity among Asian Americans*. Walnut Creek, CA: AltaMira Press.
Min, Pyong Gap (2006). "Major Issues Related to Asian American Experiences." In Pyong Gap Min (ed.), *Asian Americans: Contemporary Trends and Issues*. 2nd ed. Thousand Oaks, CA: Pine Forge Press.
Min, Pyong Gap and Rose Kim (2002). "Formation of Ethnic and Racial Identities: Narratices by Asian American Professionals." In Pyong Gap Min (ed.), *Second Generation: Ethnic Identity Among Asian Americans*. New York: Altamira Press.
Mollenkopf, John. H. (1999). "Urban Political Conflicts and Alliances: New York and Los Angeles Compared." In Charles Hirschman, Philip Kasinitz, and Josh DeWind (eds.), *The Handbook of International Migration: The American Experience*. New York: Russell Sage Foundation.
Mollenkopf, John and Mary Waters, Jennifer Holdaway, and Philip Kasinitz (2005). "The Ever Winding Path: Ethnic and Racial Diversity in the Transition to Adulthood," pp. 454–497. In Richard Settersten, Jr, Frank Furstenberg and Ruben Rumbaut (eds.), *On the Frontier of Adulthood*. Chicago: University of Chicago Press.
Möller, H. (1984). *Exodus der Kultur: Schriftsteller, Wissenschaftler und Künstler in der Emigration nach 1933*. Munich: C. H. Beck.

Morawska, Ewa (2001). "Immigrants, Transnationalism, and Ethnicization: A Comparison of This Great Wave and the Last," pp. 175–212. In G. Gerstle and J. Mollenkopf (eds.), *E Pluribus Unum? Contemporary and Historical Perspectives on Immigrant Political Incorporation.* New York: Russell Sage Foundation.

Morawska, Ewa (2003). "Immigrant Transnationalism and Assimilation: A Variety of Combinations and the Analytic Strategy It Suggests," pp. 133–176. In C. Joppke and E. Morawska (eds.), *Integrating Immigrants in Liberal Nation-States: From Post-national to Transnational.* New York: Palgrave MacMillan.

Moya, Jose C. (2005). "Immigrants and Associations: A Global and Historical Perspective." *Journal of Ethnic and Migration Studies* 31 (5): 833–864.

Muñoz-Sandoval, A. F., J. Cummins, C. G. Alvarado, and M. L. Ruef (1998). *Bilingual Verbal Ability Tests: Comprehensive Manual.* Itasca, IL: Riverside Publishing.

Museum of Jewish Heritage (2009). "Beyond Swastika and Jim Crow: Jewish Refugee Scholars at Black Colleges." Museum exhibit and film.

Nagin, D., and R. E. Tremblay (1999). "Trajectories of Boys' Physical Regression, Oppostition, and Hyperactivity on the Path to Physical Violent and Non-violent Juvenile Delinquency." *Child Development* 70 (5): 1181–1196.

National Research Council and Institute of Medicine (2002). *Community Programs to Promote Youth Development.* Committee on Community-Level Programs for Youth. Edited by Jacquelynne Eccles and Henniger Gootman. Washington D.C.: National Academy Press.

National Research Council (2004). *Engaging Schools: Fostering High School Students' Motivation to Learn.* Washington, D.C.: The National Academies Press.

Neckerman, Kathryn M., Prudence Carter, and Jennifer Lee (1999). "Segmented Assimilation and Minority Cultures of Mobility," *Ethnic and Racial Studies* 22 (6): 945–965.

New York City Public Schools (1991). "Test Scores of Recent Immigrants and Other Students, Grades 3–12."

Ngai, Mae (2004). *Impossible Subjects: Illegal Aliens and the Making of Modern America.* Princeton: Princeton University Press.

Nkurmah, Kwame (1957). *Ghana: The Autobiography of Kwame Nkrumah.* New York: Thomas Nelson & Sons.

Noam, G.G., B. M. Miller and S. Barry (2002). "Youth Development and Afterschool Time: Policy and Programming in Large Cities," pp. 19–39. In Gil G. Noam and Beth M. Miller (eds.), *Youth Development and After-school Time: A Tale of Many Cities. New Directions for Youth Development.* San Francisco: Jossey-Bass.

Obama, Barack (1995). *Dreams of My Father: A Story of Race and Inheritance.* New York: Random House.

O'Donnell, D. A., M. E. Schwab-Stone, and A. Z. Muyeed (2002). "Multidimensional Resilience in Urban Children Exposed to Community Violence." *Child Development* 73 (4): 1265–1282.

OECD (2005). *The Definition and Selection of Key Competencies: Executive Summary.* Paris, France: The Organization for Economic Cooperation and Development.

Office of Immigration Statistics (2006). *Statistical Yearbook of Immigration.* Washington, D.C.: Department of Homeland Security.

Office of Refugee Resettlement (2008). *Unaccompanied Alien Children.* Washington, D.C.: Maureen Dunn.

Ogbu, John (1991). "Low School Performance as an Adaptation: The Case of Blacks in Stockton, California," pp. 249–286. In M. Gibson and J. Ogbu (eds.), *Minority Status and Schooling: A Comparative Study of Immigrant and Involuntary Minorities.* New York: Garland.

Olsen, L. (1997). *Made in America: Immigrant Students in our Public Schools.* NY: The New Press.

Ong, Aihwa (1999). *Flexible Citizenship: The Cultural Logic of Transnationality.* Durham, NC: Duke University Press.

Orfield, G., and C. Lee (2006). *Racial Transformation and the Changing Nature of Segregation.* Cambridge, MA: The Civil Rights Project at Harvard University.

Orfield, G., and J. T. Yun (1999). *Resegregation in American Schools.* Cambridge: The Civil Rights Project, Harvard University.

ORR (Office of Refugee Resettlement) (2003). Annual Report to Congress -2003. US Department of Health and Human Services, Administration for Children and Families, Office of Refugee Resettlement. Accessed at: http://www.acf.hhs.gov/programs/orr/policy/03arc2.htm#_Ref532806959. September 12, 2006.

ORR (Office of Refugee Resettlement) (2006). Annual Report to Congress -2003 III. The Lost Boys of Sudan. US Department of Health and Human Services, Administration for Children and Families, Office of Refugee Resettlement. Accessed at: www.acf.hhs.gov/programs/orr/policy/03arc9.htm#2. September 12, 2006.

ORR (Office of Refugee Resettlement) (2007). Annual Report to Congress -2005. US Department of Health and Human Services, Administration for Children and Families, Office of Refugee Resettlement. Accessed at: http://www.acf.hhs.gov/programs/orr/data/05arc7.htm#2. September 12, 2006.

Ortiz, V. and R. Cooney (1985). "Sex-role Attitudes and Labor Force Participation Among Hispanic Females and non-Hispanic White Females," pp. 174–182. In R. O. de la Garza, F. D. Bean, C. M. Bonjean, R. Romo, and R. Alvarez (eds.), *The Mexican American Experience: An Interdisciplinary Anthology.* Austin: University of Texas Press.

Park, Robert E. (1922). *The Immigrant Press and Its Control.* New York: Harper Bros. (Republished Westport, CT: Greenwood Press).

Passel, Jeffrey S. (2006). *The Size and Characteristics of the Unauthorized Migrant Population in the U.S.* Washington, D.C.: Pew Hispanic Center.

Passel, Jeffrey S., Randolph Capps, Michael E. Fix (2004). Undocumented Immigrants: Facts and Figures (Fact Sheet/Data at a Glance). The Urban Institute. Accessed at: http://www.urban.org/url.cfm?ID=1000587. January 12, 2004.

Patrick, Erin (2004). *The US Refugee Resettlement Program. Migration Information Source.* www.migrationinformation.org/Features/print.cfm?1D-229

Patterson, Orlando (2000). "Taking Culture Seriously: A Framework and an Afro-American Illustration." In Laurence Harrison and Samuel Huntington (eds.), *Culture Matters: How Values Shape Human Progress.* New York: Basic Books.

Peck, Abraham J. (ed.) (1989). *The German-Jewish Legacy in America, 1938–1988.* Detroit, MI: Wayne State University Press.

Peréz, Lisandro (2007). "Cuba," pp. 386–398. In Mary Waters and Reed Ueda with Helen B. Marrow (eds.), *The New Americans: A Guide to Immigration Since 1965.* Cambridge: Harvard University Press.

Perlmann, Joel (1985). "Curriculum and Tracking in the Transformation of the American High School: Providence, R.I. 1880–1930," *Journal of Social History* 19 (Fall): 29–55.

Perlmann, Joel (1988). *Ethnic Differences: Schooling and Social Structure Among the Irish, Italians, Jews, and Blacks in an American City, 1880–1935.* New York: Cambridge University Press.

Perlmann, Joel (2000). "What the Jews Brought: East-European Jewish Immigration to the United States, c. 1900." In Hans Vermeulen and Joel Perlmann (eds.), *Immigrants, Schooling and Social Mobility: Does Culture Make a Difference?* New York: St. Martin's Press.

Perlmann, Joel (2005). *Italians Then, Mexicans Now: Immigrant Origins and Second-Generation Progress, 1890–2000.* New York: Russell Sage Foundation Press and the Levy Economics Institute.

Perlmann, Joel and Roger Waldinger (1997). "Second Generation Decline? Children of *Immigrants*, Past and Present—A Reconsideration", *International Migration Review* 31: 893–922.

Perlmann, Joel and Roger Waldinger (1999). "Immigrants, Past and Present: A Reconsideration," in Charles Hirschman, Philip Kasinitz, and Josh DeWind (eds.), *The Handbook of International Migration.* New York: Russell Sage Foundation.

Pianta, R. C. (1999). *Enhancing Relationships Between Children and Teachers.* Washington, D.C.: American Psychological Association.

Piore, Michael J. (1970). *Birds of Passage: Migrant Labor and Industrial Societies.* Cambridge: Cambridge University Press.

Plunkett, Scott W., and Mayra Y. Bámaca-Gómez (2003). "The Relationship Between Parenting, Accultural, and Adolescent Academics in Mexican-Origin Immigrant Families in Los Angeles." *Hispanic Journal of Behavioral Sciences* 25: 222–239.

Population Division (2004). *The Newest New Yorkers, 2000.* New York City Planning Department.

Portes, A. (1996). *The New Second Generation.* New York: Russell Sage Foundation.

Portes, Alejandro (2003). "Conclusion: Theoretical Convergencies and Empirical Evidence in the Study of Immigrant Transnationalism," *International Migration Review* 37 (3): 874–892.

Portes, Alejandro, Maria Patricia Fernandez-Kelly and William Haller (2005). "Segmented Assimilation on the Ground: The New Second Generation in Early Adulthood." *Ethnic and Racial Studies* 28: 1000–1040.

Portes, Alejandro, Luis Guarnizo, and Patricia Landolt (1999). "Introduction: Pitfalls and Promise of an Emergent Research Field." *Ethnic and Racial Studies* 22: 217–237.

Portes, Alejandro and Rubén Rumbaut (2001). *Legacies: The Story of the Immigrant Second Generation*. Berkeley: University of California Press.
Portes, Alejandro and Rubén G. Rumbaut (2006). *Immigrant America: A Portrait* (3nd Edition). Berkeley: University of California Press.
Portes, Alejandro and Min Zhou (1993). "The New Second Generation: Segmented Assimilation and its Variants." *The Annals of the American Academy of Political and Social Science* 530: 74–96.
Qian, Zhenchao and Daniel Lichter (2007). "Social Boundaries and Marital Assimilation: Interpreting Trends in Racial and Ethnic Intermarriage." *American Sociological Review* 72: 68–94.
Reddy, R., Rhodes, J., & Mulhall, P. (2003). The Influence of Teacher Support on Student Adjustment in the Middle School Years: A Latent Growth Curve Study." *Development and Psychopathology* 15: 119–138.
Reid, Ira de A. (1939). *The Negro Immigrant: His Background, Characteristics and Social Adjustment: 1899–1937*. New York: Columbia University Press.
Reitz, Jeffrey G. (2003). *Host Societies and the Reception of Immigrants*. La Jolla: Center for Comparative Immigration Studies, University of California, San Diego.
Rhodes, J. E. (2002). *Stand by Me: The Risks and Rewards of Youth Mentoring Relationships*. Cambridge, MA: Harvard University Press.
Rhodes, Jean and David DuBois (2006). "Understanding and Facilitating the Youth Mentoring Movement." *Social Policy Report*. Society for Research in Child Development. Accessed at: www.srcd.org/documents/publications/spr/spr20-3.pdf
Rhodes, Jean (2002). *Stand By Me: The Risks and Rewards of Mentoring Today's Youth*. Cambridge: Harvard University Press.
Riis, Jacob A. (1890). *How the Other Half Lives*. New York: Charles Scribner's.
Ritterband, Paul (1997). "Jewish Identity among Russian Immigrants in the U.S," pp. 325–343. In Noah Lewin-Epstein, Yaacov Ro'i and Paul Ritterband (eds.), *Russian Jews on Three Continents: Migration and Resettlement*. London: Frank Cass.
Rivera-Batiz, Francisco (2004). "NewYorktitlan: The Socioeconomic Status of Mexican New Yorkers". *Regional Labor Review* (Winter/Spring): 32–44.
Robben, A. C. G. M., and M. M. Suárez-Orozco (eds.) (2000). *Cultures under Siege: Collective Violence and Trauma*. Cambridge: Cambridge University Press.
Rodriguez, Richard (2002). *Brown: The Last Discovery of America*. New York: Viking.
Roediger, David (2005). *Working Toward Whiteness: How America's Immigrants Became White; The Strange Journey from Ellis Island to the Suburbs*. New York: Basic Books.
Roedinger, David R. (1991). *The Wages of Whiteness: Race and the Making of the American Working Class*. New York: Verso.
Roeser, R. W., and J. S. Eccles (1998). "Adolescents' Perception of Middle School: Relation to Longitudinal Changes in Academic and Psychological Adjustment." *Journal of Research on Adolescence* 8 (1): 123–158.
Roffman, Jennifer, Carola Suárez-Orozco, and Jean Rhodes (2003). "Facilitating Positive Development in Immigrant Youth: The Role of Mentors and

Community Organizations." In Francisco Villareal, Daniel Perkins, Lynne Borden, and Joanne Keith (eds.), *Community Youth Development*. Thousand Oaks, CA: Sage Publications.

Rong, Xue Lan and Frank Brown (2001). "The Effects of Immigrant Generation and Ethnicity on Educational Attainment among Young African and Caribbean Blacks in the United States." *Harvard Educational Review* 71 (3): 536–565.

Ruiz-de-Valasco, J., M. Fix, and B. C. Clewell (2001). *Overlooked and Underserved: Immigrant Students in U.S. Secondary Schools*. Washington, DC: The Urban Institute.

Rumbaut, R. G., and A. Portes (2001). *Ethnicities: Children of Immigrants in America*. New York: Russell Sage Foundation.

Rumbaut, Rubén G. (2007). "Vietnam," pp. 652–673. In Mary Waters and Reed Ueda with Helen B. Marrow (eds.), *The New Americans: A Guide to Immigration Since 1965*. Cambridge, Harvard University Press.

Rumbaut, Ruben (1999). Assimilation and Its Discontents: Ironies and Paradoxes." In Charles Hirschman, Philip Kasinitz and Josh DeWind (eds), *The Handbook of International Immigration: The American Experience*. New York: Russell Sage Foundation.

Sabagh, Georges, and Mehdi Bozorgmehr (2003). "From 'Give Me your Poor' to 'Save Our State': New York and Los Angeles as Immigrant Cities and Regions," pp. 99–123. In David Halle (ed.), *New York & Los Angeles: Politics, Society, and Culture, A Comparative View*. Chicago: University of Chicago Press.

Saenger, G. (1941). *Today's Refugees, Tomorrow's Citizens: A Story of Americanization*. New York: Harper & Brothers.

Salamon, Lester M. (1999). *America's Nonprofit Sector: A Primer*. New York: Foundation Center.

Salamon, Lester M. (2002). *The State of Nonprofit America*. Washington: Brookings Institution Press.

Salamon, Lester M. (2006). *The Tools of Government: A Guide to the New Governance*. New York: Oxford University Press.

Salvo, Joseph (2005). Mexicans and the Census in New York. Presentation at Baruch College Conference on The Mexican Population in New York City. May 13.

Samdal, O., D. Nutbeam, B. Wold, and L. Kannas (1998). "Achieving Health and Educational Goals Through Schools: A Study of the Importance of School Climate and the Student's Satisfaction with School." *Health Education Research* 13 (3): 383–397.

Sarason, I. G., B. R. Sarason, and G. R. Pierce (1990). "Social Support: The Search for Theory." *Journal of Social and Clinical Psychology* 9: 133–147.

Sarroub, L. K. (2001). "The Sojourner Experience of Yemeni American High School Students: An Ethnographic Portrait." *Harvard Educational Review* 71 (3): 390–415.

Saxton, Alexander (1971). *The Indispensable Enemy: Labor and the Anti-Chinese Movement in California*. Berkeley: University of California Press.

Schrover, Marlou and Floris Vermeulen (2005). "Immigrant Organizations." *Journal of Ethnic and Migration Studies* 31 (5): 823–832.

Schlesinger, Arthur M. Jr. (1986). *The Cycles of American History.* Boston: Houghton Mifflin.
Schuck, Peter H. (1998). *Citizens, Strangers, and In-Betweens: Essays on Immigration and Citizenship.* Boulder, CO: Westview Press.
Schunk, D. H. (1991). "Self-efficacy and Academic Motivation." *Educational Psychologist* 26: 207–231.
Sirin, S. R., and L. Rogers-Sirin (2005). "Components of School Engagement Among African American Adolescents." *Applied Developmental Science* 9 (10): 5–13.
Skrentny, John (2002). *The Minority-Rights Revolution.* Cambridge: Harvard University Press.
Slotnik, W. J., and D. B. Gratz (1999). "Guiding Improvement." *Thrust for Educational Leadership* 28 (3): 10–12.
Smith, Malcolm and Dee Brewer (n.d.). "Stop Summer Academic Loss—An Educational Policy Priority." *A MetaMetrics Paper.* Available at www.lexile.com.
Smith, Robert Courtney (2006). *Mexican New York: Transnational Worlds of New Immigrants.* Berkeley: University of California Press.
Smith, Robert Courtney (2002). "Race, Ethnicity and Gender in the School Outcomes of Second Generation Mexican Americans in New York," pp. 110–125. In Marcelo Suárez-Orozco and Mariela Paez (eds.), *Latinos in the 21st Century.* Berkeley: University of California Press.
Smith, Robert C. (2008). "Horatio Alger Lives in Brooklyn." In Alejandro Portes and M. Patricia Fernandez Kelly (eds.), *Exceptional Outcomes: Achievement in Education and Employment Among Children of Immigrants.* The ANNALS of the American Academy of Political and Social Science Series.
Sokal, M. M. (1984). "The Gestalt Psychologists in Behaviorist America". *American Historical Review* 89: 1240–1263.
Solomon, Barbara Miller (1956). *Ancestors and Immigrants: A Changing New England Tradition.* Cambridge, MA: Harvard University Press.
Sonnert, Gerhard, and Gerald Holton (2002). *Ivory Bridges: Connecting Science and Society.* Cambridge, MA: MIT Press.
Sonnert, Gerhard and Gerald Holton (2006). *What Happened to the Children Who Fled Nazi Persecution?* New York: Palgrave Macmillan.
South, Scott J., Kyle Crowder, and Erick Chavez (2005). "Migration and Spatial Assimilation Among U.S. Latinos: Classical Versus Segmented Trajectories." *Demography* 42: 497–521.
Soysal, Tasemin Nuhoglu (1994). *Limits of Citizenship: Migrants and Postnational Membership in Europe.* Chicago: University of Chicago Press.
Spaulding, E. W. (1968). *The Quiet Invaders: The Story of the Austrian Impact Upon America.* Vienna: Österreichischer Bundesverlag für Unterricht, Wissenschaft und Kunst.
Spencer, Herbert (1899). *The Principles of Sociology,* Vol. II. New York: D. Appleton and Co.
Stadler, F. (ed.) (1987). *Vertriebene Vernunft I: Emigration und Exil österreichischer Wissenschaft 1930–1940.* Vienna: Jugend und Volk.

Stadler, F. (ed.) (1988). *Vertriebene Vernunft II: Emigration und Exil österreichischer Wissenschaft 1930–1940*. Vienna: Jugend und Volk.
Stanton-Salazar, R. S. (2004). "Social Capital Among Working-class Minority Students". In M. A. Gibson, P. Gándara & J. P. Koyma (eds.), *School Connections: U.S. Mexican Youth, Peer, & School Achievement*. New York: Teacher's College Press.
Steinberg, S., B. B. Brown, and S. M. Dornbusch (1996). *Beyond the Classroom*. New York: Simon and Schuster.
Strauss, H. A., K. Fischer, C. Hoffmann, and A. Söllner (eds.) (1991). *Die Emigration der Wissenschaften nach 1933: Disziplingeschichtliche Studien*. Munich: K. G. Saur.
Suárez-Orozco, Carola, and Desirée Baolian Qin (2006). "Gendered Perspectives in Psychology: Immigrant Origin Youth." *International Migration Review* 40: 165–198.
Suárez-Orozco, C. and D. B. Qin-Hillard (2004). "The Cultural Psychology of Academic Engagement: Immigrant Boys' Experiences in U.S. Schools." In Way, Niobe & Judy Chu (eds), *Adolescent Boys: Exploring Diverse Cultures of Boyhood*. New York: New York University Press.
Suárez-Orozco, C., and M. M. Suárez-Orozco (1995). *Transformations: Immigration, Family Life, and Achievement Motivation Among Latino Adolescents*. Stanford CA: Stanford University Press.
Suárez-Orozco, C., and M. M. Suárez-Orozco (2001). *Children of Immigration*. Cambridge, MA: Harvard University Press.
Suárez-Orozco, C., M. Suárez-Orozco and I. Todorova (2007). *Moving Stories: Academic Passages of Immigrant Youth*. Cambridge, MA: Harvard University Press.
Suárez-Orozco, C., M. Suárez-Orozco, and I. Todorova (2008). *Learning a New Land: Immigrant Students in American Society*. Cambridge, MA: Harvard University Press.
Suárez-Orozco, Carola, Irina Todorova and Desiree Baolin Qin (2006). "The Well Being of Immigrant Adolescents: A Longitudinal Perspective on Risk and Protective Factors." In Hiram Fitzgerald, Robert Zucker and Kristine Freeark (eds.), *The Crisis in Youth Mental Health*. Westport CT: Praeger.
Suárez-Orozco, Marcelo (1991). "Immigrant Adaptation to Schooling: A Hispanic Case," pp. 37–61. In M. Gibson and J. Ogbu (eds.), *Minority Status and Schooling: A Comparative Study of Immigrant and Involuntary Minorities*. New York: Garland.
Suárez-Orozco, M., and D. B. Qin-Hilliard (2004). *Globalization: Culture and Education in the New Millennium*. Berkeley, CA: University of California Press.
Sumner, William Graham (1911). *Folkways: A Study of the Sociological Importance of Usages, Manners, Customs, Mores, and Morals*. Boston: Ginn and Co., 1906.
Sung, Betty Lee (1987). *The Adjustment of Chinese Immigrant Children in New York City*. New York: Center for Migration Studies.
Teachman, Jay D. (1987). "Family Background, Educational Resources, and Educational Attainment." *American Sociological Review* 52: 548–557.

Thernstrom, Stephan (1973). *The Other Bostonians: Poverty and Progress in the American Metropolis, 1880–1970.* Cambridge, MA: Harvard University Press.

Tyack, David B. (1974). *The One Best System: A History of American Urban Education.* Cambridge, MA: Harvard University Press.

U.S. Bureau of the Census (2006). *American Community Survey for Metropolitan Los Angeles.* http://www.census.gov/acs/www/.

U.S. Commission on Immigration Reform (1994). *U.S. Immigration Policy: Restoring Credibility.* Washington, D.C.: U.S. Government Printing Office.

U.S. Department of Education (1994). *Research Report: What Do Student Grades Mean? Differences Across Schools*: Office of Research.

U.S. Department of Education (2006). *Executive Summary of the No Child Left Behind Act of 2001.* Retrieved in October 2006 from http://www.ed.gov/nclb/overview/intro/ execsumm.pdf

United Nations High Commissioner on Refugees (2006, March). *Background Paper on the Identification, Referral, and Reception of Asylum-Seeking Separated Children.* Washington, D.C.: UNHCR.

United States Bureau of the Census (2000). Census of Population, 5% Public Use Microsample.

Valenzuela, A. (1999). "Gender Roles and Settlement Activities Among Children and Their Immigrant Families." *American Behavioral Scientist* 42 (4): 720–742.

Van Hook, Jennifer, Susan K. Brown, and Frank D. Bean (2006). "For Love or Money? Welfare Reform and Immigrant Naturalization." *Social Forces* 85 (2): 643–666.

Vera Institute of Justice (2009). *Unaccompanied Children Pro Bono Pilot Project Report.* New York.

Vermeulen, Hans (2000). "Introduction: The Role of Culture in Explanations of Mobility." In Hans Vermeulen and Joel Perlmann (eds.), *Immigrants, Schooling and Social Mobility: Does Culture Make a Difference?* New York: St. Martin's Press.

Vermeulen, Hans and Tijno Venema (2000). "Peasantry and Trading Diaspora: Differential Mobility of Italians and Greeks in the United States." In Hans Vermeulen and Joel Perlmann (eds.), *Immigrants, Schooling and Social Mobility: Does Culture Make a Difference?* New York: St. Martin's Press.

Vickerman, Milton (1999). *Crosscurrents: West Indian Immigrants and Race.* New York: Oxford University Press.

Villarrubia, Jacqueline, Nancy Denton, and Richard Alba (2007). "Gateway State, Not Gateway City: New Immigrants in the Hudson Valley," presented at the annual meeting of the Population Association of America.

Waddell, S. (2001/2002). "Societal Learning: Creating Big-Systems Change," *The Systems Thinker* 12 (10, December–January): 1, Pegasus Communication.

Wagmiller, Robert L., Jr., Mary Clare Lennon, Li Kuang, Philip M. Alberti, and J. Lawrence Aber (2006). "The Dynamics of Economic Disadvantage and Children's Life Chances." *American Sociological Review* 71: 847–866.

Waldinger, Roger (1996). *Still the Promised City? African-Americans and New Immigrants in Postindustrial New York.* Cambridge: Harvard University Press.

Waldinger, R., and C. Der-Martirosian (2001). "The Immigrant Niche: Pervasive, Persistent, Diverse," pp. 228–271. In Waldinger, R. (ed.), *Strangers at the Gates:*

New Immigrants in Urban America, Berkeley, CA: University of California Press.

Waldinger, Roger and David Fitzgerald (2004). "Transnationalism in Question" (with David Fitzgerald) *American Journal of Sociology* 109 (5): 1177–1195.

Waldinger, Roger and Michael Lichter (2003). *How the Other Half Works: Immigration and the Social Organization of Labor.* Berkeley: University of California.

Warner, William Lloyd, and Leo Srole (1945). *The Social Systems of American Ethnic Groups.* New Haven: Yale University Press.

Waters, Mary (1996). "The Intersection of Gender, Race, and Ethnicity Development of Caribbean American Teens." In B. J. Leadbeater & N. Way (eds.), *Urban Girls: Resisting Stereotypes, Creating Identities.* New York: New York University Press.

Waters, Mary (1999). *Black Identities: West Indian Immigrant Dreams and American Realities.* Cambridge, MA: Harvard University Press.

Wenger, Beth (1996). *New York Jews and the Great Depression.* New Haven: Yale University Press.

Wentzel, K. R. (1999). "Social Influences and School Adjustment: Commentary." *Educational Psychologist* 34 (1): 59–69.

White, K. (1982). "The Relationship Between Socio-economic Status and Academic Achievment." *Psychological Bulletin* 91: 461–481.

Williams, L. Susan, Sandra D. Alvarez, and Kevin s. Andrade Hauck (2002). "My Name Is Not María: Young Latinas Seeking Home in the Heartland." *Social Problems* 49: 563–584.

Wills, T. A. (1985). "Supportive Functions of Interpersonal Relationships," pp. 61–82. In S. L. S. S. Cohen (ed.), *Social Support and Health.* Orlando: Academic Press.

Wilson, W. (1997). *When Work Disappears: The World of the New Urban Poor.* New York: Vintage Books.

Wilson. William Julius (1977). *The Declining Significance of Race.* Chicago: University of Chicago Press.

Yang, Philip Q. (1994). "Explaining Immigrant Naturalization." *International Migration Review* 28: 449–477.

Zeltzer-Zubida, Aviva (2004). "Affinities and Affiliations: The Many Ways of Becoming a Russian Jewish American," pp. 339–360. In Philip Kasinitz, John Mollenkopf and Mary Waters (eds.), *Becoming New Yorkers: The Second Generation in a Global City.* New York: The Russell Sage Foundation Press.

Zeltzer-Zubida, Aviva and Philip Kasinitz (2005). "The Next Generation: Russian Jewish Young Adults in Contemporary New York." *Contemporary Jewry* 25: 193–225.

Zhou, Min (1997a). "Social Capital in Chinatown: The Role of Community-Based Organizations and Families in the Adaptation of the Younger Generation." In Maxine Seller and Lois Weis (eds.), *Beyond Black and White: New Faces and Voices in U.S. Schools.* Albany: State University of New York Press.

Zhou, Min (1997b). "Growing Up American: The Challenge Confronting Immigrant Children and Children of Immigrants." *Annual Review of Sociology* 23: 69–95.

Zhou, Min (2001). "Progress, Decline, Stagnation? The New Second Generation Comes of Age." In Roger Waldinger (ed.), *Strangers at the Gates: New Immigrants in Urban America*. Berkeley: University of California Press.

Zhou, Min (2009). *Contemporary Chinese America: Immigration, Ethnicity, and Community Transformation*. Philadelphia: Temple University Press.

Zhou, Min and Carl L. Bankston, III. (1998). *Growing Up American: How Vietnamese Children Adapt to Life in the United States*. New York: Russell Sage Foundation.

Zhou, Min and Susan Kim (2006). "Community Forces, Social Capital, and Educational Achievement: The Case of Supplementary Education in the Chinese and Korean Immigrant Communities." *Harvard Educational Review* 76: 1–29.

Zuniga, Victor and Ruben Hernandez (2005). *New Destinations for Mexican Migration*. New York: Russell Sage Foundation.

Contributors

Clark C. Abt is founder, chairman emeritus, and past president of Abt Associates Inc. Dr. Abt was born in Cologne, Germany, in 1929; came to the United States in 1937; and served in the U.S. Air Force. Dr. Abt has a B.S. in industrial engineering and a Ph.D. in political science from MIT, and has taught at Boston University, Brandeis University, Cambridge College, Franklin Institute of Technology, Harvard University, Johns Hopkins University, and the University of Massachusetts. At B.U., from 1991 to 1994, Dr. Abt organized and directed four annual Russian-American Entrepreneurial Workshops in Defense Technology Conversion for Russian and U.S. nuclear weapons scientists. He has written and/or edited ten books, including *Serious Games (1970), The Social Audit for Management (1976), The Evaluation of Social Programs (ed.) (1977), The Termination of a Nuclear War (1985)*. For the last ten years, he has also been a volunteer part-time tutor in Boston inner-city public high schools. Since September 11, 2001, until 2007, he devoted much of his research and writing to homeland defense of port cities against catastrophic nuclear and biological terrorism, including a graduate course on *Terrorism and Disaster Management in Health Care Settings* he created and taught as Distinguished Professor of Management at Cambridge College. From 2007 to the present (2010), most of his teaching and research as Professor of Energy & Sustainable International Development at Brandeis University has been devoted to the benefit-cost analysis of renewable energy technologies (wind, solar, geothermal, tidal). He is a member of the Board of Directors of Economists for Peace & Security and Physicians for Social Responsibility of New England.

The seeds of **Richard Alba**'s interest in race and ethnicity were sown during his childhood in the Bronx of the 1940s and 1950s and nurtured intellectually at Columbia University, where he received his undergraduate and graduate education, completing his Ph.D. in 1974. He is currently Distinguished Professor of Sociology at the Graduate Center of the City University of New York. Besides race/ethnicity, his teaching and research focus on international migration in the United States and in Europe, and he has done research in France and in Germany, with the support of Fulbright grants and fellowships from the Guggenheim Foundation, the German Marshall Fund, and Russell Sage Foundation. His books include *Ethnic Identity: The Transformation of White America* (1990); *Italian Americans: Into the Twilight of Ethnicity* (1985); and *Remaking the American Mainstream: Assimilation and Contemporary Immigration* (2003), cowritten with Victor Nee. His most recent book is *Blurring the Color Line: The New Chance for a More Integrated America*, published in

2009. Dr. Alba has been president of the Eastern Sociological Society (1997–1998) and vice president of the American Sociological Association (2000–2001).

James Bachmeier is a postdoctoral student in sociology at the University of California, Irvine. His research interests include the sociology of international migration and immigrant group incorporation, racial/ethnic stratification in the U.S. labor market, and labor force participation among youth. His research has received numerous awards, including the A. Kimball Romney Award for the best graduate student paper in the School of Social Sciences at UC-Irvine, and Order of Merit Award, in 2008.

Frank D. Bean is Chancellor's Professor of Sociology and Economics and director of the Center for Research on Immigration, Population and Public Policy at the University of California, Irvine. His most recent books are *Help or Hindrance? The Economic Implications of Immigration for African Americans* (coedited with Daniel Hamermesh), *Immigration and Opportunity* (coedited with Stephanie Bell-Rose), *America's Newcomers and the Dynamics of Diversity* (with Gillian Stevens), and *The Diversity Paradox: Immigration and the Color Line in 21st Century America* (with Jennifer Lee), all published by the Russell Sage Foundation. His current research and writing focus on the determinants and consequences of U.S. immigration patterns and policies, immigrant group (especially Mexican origin) integration, immigrant naturalization and integration, estimates of U.S. immigration and emigration, the implications of immigration for changing patterns and definitions of race/ethnicity in the United States, and world population trends and policies.

Susan K. Brown is associate professor of sociology at the University of California, Irvine. She is the author of *Beyond the Immigrant Enclave: Network Change and Assimilation*. Her research focuses on immigrants' incorporation, particularly residential integration.

Robert J. Carey is vice president for Resettlement and Migration Policy at the International Rescue Committee and Chair of the Refugee Council USA. He currently is part of the Department of State's Joint Regional Working Group for Southeast Asia. Mr. Carey frequently speaks to the media about resettlement, refugee admissions, and U.S. immigration policy. He has appeared on the BBC, CNN, MSNBC, and NPR, and has been interviewed by the *New York Times*, the *Wall Street Journal*, and the *Boston Globe*, among other publications.

Julianne S. Duncan, Ph.D., is associate director for children's services, Office of Refugee Programs, Migration and Refugee Services (MRS), United States Conference of Catholic Bishops (USCCB) in Washington, D.C. In this capacity, she manages programs for refugee children, asylees, undocumented alien minors, and trafficking victims. Her responsibilities have included oversight of the USCCB/MRS field coordination services that assist undocumented children. During 2002–2003, she assisted the Office of Refugee Resettlement to design and implement that program. She is an internationally recognized expert in the child welfare issues of migrant children. During the year 2000, she worked in Kakuma Refugee Camp in Kenya, assisting in the resettlement preparation for the Sudanese children and young adults who have been resettled in the United States. She has worked in refugee child welfare and mental health programs in Washington State for Lutheran Social Services, Catholic Community Services, and the State of Washington. Her earlier work involved assistance to Southeast Asian refugees in Thailand and in the United States. She has developed curriculum for parenting classes for Cambodian and other refugee ethnic groups and taught parenting classes for refugees for many years. Dr. Duncan's degree is in anthropology with a specialization in the mental health needs of refugees. She is a licensed mental health counselor with certification in child and minority mental health.

Nancy Foner is Distinguished Professor of Sociology at Hunter College and the Graduate Center of the City University of New York. Her recent work on immigration has compared U.S. immigration today and a century ago, the immigrant experience in various U.S. gateway cities, and immigrant minorities in the United States and Europe. Her many books include *In a New Land: A Comparative View of Immigration* (NYU Press, 2005), *From Ellis Island to JFK: New York's Two Great Waves of Immigration* (Yale University Press and Russell Sage Foundation, 2000), and (as editor) *Across Generations: Immigrant Families in America* (NYU Press, 2009) and Not *Just Black and White: Historical and Contemporary Perspectives on Immigration, Race, and Ethnicity in the United States* (Russell Sage Foundation, 2004, with George Fredrickson).

Francisco X. Gaytán is assistant professor of social work at Northeastern Illinois University in Chicago. His research focuses on the socio-emotional and academic development of first and second generation immigrant youth in the United States. More specifically, he has examined the role of social support and social capital on the cultural, academic, and psychological adaptation of the growing Mexican immigrant student population in New York City.

Nathan Glazer is professor emeritus of sociology and education at Harvard University. He is the author of *American Judaism* (University of Chicago Press, 1957), revised a few times since and still in print, and of books on American ethnicity and social policy, among them *Beyond the Melting Pot* (with Daniel P. Moynihan), *Affirmative Discrimination, Ethnic Dilemmas, The Limits of Social Policy*, and *We are All Multiculturalists Now*, the latter four with Harvard University Press. He was one of the founding editors of the magazine *Commentary*, and served for 30 years as coeditor of the journal the *Public Interest*. His most recent book, *From a Cause to a Style*, on modernist architecture and the city, was published by Princeton University Press in the Spring of 2007.

Steven J. Gold is professor and graduate program director in the Department of Sociology at Michigan State University. His research involves ethnic economies, ethnic communities, and international migration. He is coeditor of *Immigration Research for a New Century: Multidisciplinary Perspectives* (with Rubén G. Rumbaut and Nancy Foner) and the author *of Refugee Communities: A Comparative Field Study, From the Worker's State to the Golden State, Ethnic Economies* (with Ivan Light), and *The Israeli Diaspora*, which won the Thomas & Znaniecki Book Award in 2003. His book, *The Store in the Hood: A Century of Ethnic Business and Conflict* was published by Roman and Littlefield in 2010.

Marilyn Halter is professor of history at Boston University where she is also a research associate at the Institute on Culture, Religion and World Affairs. Her books include *Between Race and Ethnicity: Cape Verdean American Immigrants, 1860–1965* (1993); *Shopping for Identity: The Marketing of Ethnicity* (2000); and her edited collection, *New Migrants in the Marketplace: Boston's Ethnic Entrepreneurs* (1995). Halter's current research project, coauthored with Violet Johnson, is "African and American," a study of issues of identity formation and socioeconomic incorporation among recent West African immigrants and refugees to the United States.

Gerald Holton is Mallinckrodt Research Professor of Physics and research professor of history of science at Harvard University. His main interests are in the history of modern physics, the physical properties of matter under high pressure, and the careers of young persons. He is a fellow of the usual list of organizations in the United States and abroad, and served as president of the History of Science Society and on several U.S. national commissions. His book publications include *Thematic Origins of Scientific Thought; Science and Anti-Science; Einstein, History, and Other Passions; Victory and Vexation in Science*; and several books with Gerhard Sonnert (see his bio).

Violet M. Showers Johnson is professor of history and director of the Women's Global Leadership Center at Agnes Scott College, where she teaches courses on race, ethnicity, and immigration; African American history; and the history of the African Diaspora. Her scholarly work focuses on black immigrants in the United States. She is author of *The Other Black Bostonians: West Indians in Boston, 1900–1950* (Indiana University Press, 2006). Johnson is currently collaborating with Marilyn Halter on the research project, "African and American."

Philip Kasinitz is professor of sociology at Hunter College and the Graduate Center of the City University of New York. He is the former president of the Eastern Sociological Society and the former chair of the International Migration Section of the American Sociological Association. With Mary Waters and John Mollenkopf, he directs the Second Generation in Metropolitan New York research project, funded by the Russell Sage Foundation.

Jane S. Kim is technical advisor, youth programs, with the International Rescue Committee. Since joining the IRC in 1998, she has served in various capacities, including refugee caseworker based in Croatia, national immigration legal officer, and director of anti-trafficking initiatives. In addition, she possesses extensive child protection experience and has recently worked in Liberia and Cambodia, as well as with Human Rights Watch and the Legal Aid Society Juvenile Rights Division in previous years. She has earned a joint JD/MSW degree from the University of Pennsylvania and a BA degree in child development from Tufts University.

Mark A. Leach is assistant professor of rural sociology and demography at the Pennsylvania State University. His research interests include explaining Mexican migrants' increased settlement in rural and urban nontraditional destinations in the United States during the 1990s and understanding related change in migrants' economic outcomes. He is also currently investigating processes of Mexican immigrant household formation, their implications for the wellbeing of the children of immigrants, and whether and how such processes differ across places.

Joel Perlmann received his Ph.D. in history and sociology at Harvard. He is a senior scholar at the Levy Economics Institute of Bard College and a research professor at the college. Among his publications are *Ethnic Differences: Schooling and Social Structure among the Irish, Italians, Jews, and blacks in an American City, 1880–1935* and *Italians Then, Mexicans Now: Immigrant Origins and Second-Generation Progress, 1890–2000*.

Robert Smith is professor of sociology, migration studies and public affairs at the Baruch College School of Public Affairs and the Graduate Center, City University of New York. He is the author of Mexican New York: Transnational Worlds of New Immigrants (California, 2006), which won three section awards and the Distinguished Scholarly Publication Award from the American Sociological Association in 2008. During 2009–2010, he is a John Simon Guggenheim Fellow. He is the cofounder of the Mexican Educational Foundation, a nonprofit organization promoting education and leadership in the Mexican community.

Gerhard Sonnert is an associate of the physics department at Harvard University and a research associate in the science education department at the Harvard-Smithsonian Center for Astrophysics. A sociologist by training, he studied, with Gerald Holton, young refugees from Central Europe (*What Happened to the Children Who Fled Nazi Persecution*, 2006). Other interests include women in science (*Who Succeeds in Science? The Gender Dimension*, 1995, and *Gender Differences in Science Careers: The Project Access Study*, 1995, both with G. Holton); science policy (*Ivory Bridges: Connecting Science and Society*, 2002, with G. Holton); and history of science (*Einstein and Culture*, 2005).

Carola Suárez-Orozco, professor of applied psychology at NYU. Research foci include the intersection of cultural and psychological factors in the adaptation of immigrant and ethnic minority youth, academic engagement, the role of the "social mirror" in identity formation, immigrant family separations, the role of mentors in facilitating positive development in immigrant youth, and the gendered experiences of immigrant youth. Coauthor of *Learning a New Land: Immigrant Children in American Society* (Harvard University Press, 2001) *Children of Immigration* (Harvard University Press, 2001) and *Transformations: Migration, Family Life, and Achievement Motivation Among Latino Adolescents* (Stanford University Press,

1995) Co-editor of the six-volume *Interdisciplinary Perspectives on The New Immigration* (Routledge Press, 2001) and *The New Immigration: An Interdisciplinary Reader* (Routledge Press, 2005).

Ken Tota is the current Deputy Director for the Office of Refugee Resettlement (ORR), Administration for Children and Families (ACF), U.S. Department of Health and Human Services (HHS). Prior to ORR, he was directly involved in overseeing the transfer of the Unaccompanied Alien Children's Program to ORR from the Office of Juvenile Affairs, Immigration and Naturalization Service as a result of the Homeland Security Act. In September 2006, Ken presented at an international roundtable hosted by the U.S. Embassy in Greece and the Greek Council for Refugees to discuss options for improving the treatment and care of unaccompanied children in Greece. Before entering civil service, Ken worked for the United States Conference of Catholic Bishops (USCCB) in Washington, D.C. and Miami, FL, where he coordinated resettlement efforts to a series of mass migrations from both Cuba and Haiti. Ken received a master's in public administration from the American University in Washington, D.C.

Hui-shien Tsao obtained her Ph.D. in sociology from the University at Albany, SUNY, and is the manager of user support in Albany's Center for Social and Demographic Analysis. In addition to offering computing and statistical expertise to faculty and graduate student patrons at the center, she conducts a research program of her own. Originally interested in work and occupations, her research emphases have been influenced and shaped by a group of demographers and urban sociologists at Albany. Her interests reflect these research areas with a focus on the adaptation of immigrants, often concentrating on residential and work-related experiences, internal migration, and the impact of government programs on urban communities. She has published in a variety of scholarly journals including the *Sociological Quarterly, Sociological Spectrum, Journal of Urban Affairs, and Urban Studies*. A recent article, written with Deirdre Oakley, appeared in *Cities*.

Reed Ueda is Professor of History at Tufts University. He is engaged in the study of international migration from a comparative historical perspective.

Roger Waldinger is Distinguished Professor of sociology, UCLA. He is the author of several books, including *Strangers at the Gates: New Immigrants in Urban America* (University of California Press, 2001) and *How the Other Half Works: Immigration and the Social Organization of Labor* (coauthored with Michael Lichter; University of California Press, 2003). He is also the son of two "second wave" immigrants.

Mary C. Waters is the M.E. Zukerman Professor of Sociology at Harvard University. She specializes in the study of immigration, intergroup relations, the formation of racial and ethnic identity among the children of immigrants, and the challenges of measuring race and ethnicity. Her most recent books are *The New Americans: A Guide to Immigration Since 1965* (co-edited with Reed Ueda and Helen Marrow, Harvard University Press, 2007) and *Inheriting the City: The Children of Immigrants*

Come of Age (co-authored with Philip Kasinitz, John H. Mollenkopf and Jennifer Holdaway, Harvard University Press, 2008), which won the American Sociological Association's Distinguished Contribution to Scholarship award in 2010.

Annie Wilson is the executive vice president of Lutheran Immigration and Refugee Service. She serves as the chief operating officer for the organization, overseeing service and advocacy programs for refugees, immigrants, asylum seekers, people in immigration detention, and children and youth. She attended Williams College and graduated with a major in religion. She has also completed a nonprofit management program at Johns Hopkins University.

Index

Abt, Clark, 35
academic performance, 40, 117,
    151–65, 171, 180–4, 193–9, 204–5
    see also education
Adams, Henry, 28
Addams, Jane, 24
Adelante Alliance's Distance Learning
    Program, 183
Adorno, Theodor, 56
affirmative action, 74, 93, 101, 107–11,
    251, 268–71
African Abroad (journal), 118
African Americans, 90–1, 93, 98–9,
    105–10, 114, 118, 120–4, 172, 179,
    196, 267, 269–70, 276–8
Alba, Richard, 73, 94, 251, 254
Alvarez, Richard, 188n4
Anglo-conformity, 97, 102
Anschluss (annexation), 39
anti-Semitism, 41, 55
Arendt, Hannah, 37, 56–7
assimilation, 1, 4, 27–8, 31–2, 42–4,
    54–9, 73, 94–6, 102, 118, 131, 178,
    254, 258, 268–70, 276
    downward, 24, 30, 73, 89, 111
    segmented, 25, 94, 172–3, 182, 267
Associacion Tepeyac, 187
asylum, 62–3, 67–8, 189, 220, 223,
    226–30, 240, 244, 249
attached minors, 222, 224n8
Azikiwe, Nnamdi, 115

Bachmeier, James, 74
Baruch College, 183, 185–7
Bean, Frank D., 74, 130–1
Benjamin, Walter, 55–6

Berlin, Irving, 95
Bethe, Hans, 37
Biafran War, 115
Bility, Hassan, 67
Birdsell, David, 185, 188n4
Black Mexicans, 109
Black Panthers, 110
Bourdieu, Pierre, 17
Boyd, Monica, 97
Bridging Refugee Youth and
    Children's Services (BRYCS), 165,
    213, 218
Brown, Susan K., 74, 130
Buber, Martin, 55
Burr, Clinton Stoddard, 29
Bush, George W., 171

Cape Verdean Alumni Network
    (CVAN), 124
Carey, Robert J., 189
Castles, Stephen, 88
Castro, Fidel, 211
Catholic Bishops, 189, 209–22, 224n2,
    228
citizenship, 130–4, 139–40, 144–6, 226,
    244, 263
City University of New York (CUNY),
    99, 103–4, 149, 177–88, 188n2
civic pluralism, 102
civil rights movement, American, 74,
    93, 106–10, 115, 121, 276
Cleveland, Grover, 25
Cohen, Miriam, 14
Cold War, 73, 75–9, 86–8
Comparative Entrepreneurship and
    Immigration Project, 255–6

concrete knowledge, 178, 183, 185, 188n1
Congress of Mexican Community Organizations for Education (COMEXCOE), 187–8
Covello, Leonard, 97
creativity, 61, 94–7
Crook, David, 188n2
Cubberley, Ellwood P., 29
cultural capital, 17, 43–6, 48, 66, 70–2
culture shock, 43, 76

de-skilling of employment, 25
deSantillana, Giorgio, 66
Deutsch, Felix, 37
Deutsch, Helene, 37
Diallo, Amadou, 123
discrimination, 12, 19, 28–9, 63, 67, 76, 79, 83, 90–3, 97–9, 107–11, 156–7, 183, 195–6, 239, 276
diversity, institutionalization of, 106–11
Donnelly, Ignatius, 25
downward assimilation, 73, 89, 111
  see also assimilation
Du Bois, W.E.B., 172
Duncan, Julianne, 189

earnings, see income
education, 10–20, 26, 39–48, 52–8, 65, 79–92, 99–111, 115–17, 124, 138, 140–46
  see also academic performance; City University of New York (CUNY); high school graduation rates; Historically Black Colleges and Universities (HBCUs); No Child Left Behind (NCLB)
Einstein, Albert, 37, 57
English as a Second Language/English Language Learners, ESL/ELL, 195, 199, 203
  see also language acquisition
entrepreneurs and entrepreneurialism, 61–72, 75, 82–3, 181
Erikson, Erik, 37

Feinstein, Diane, 246
first generation immigrants, definition of, 257
First Wave of refugees, definition of, 38, 44
Flores Settlement Agreement, 238–9, 242, 245
Flores, O., 292
Foner, Nancy, 7, 21n3
Fredrickson, George, 11
Fromm, Erich, 57
Fuchs, Lawrence, 97

Gaytán, Francisco X., 149
Glazer, Nathan, 35, 97, 102
Glick-Schiller, Nina, 262
globalization, 29–30, 96, 225, 251, 253
Gödel, Kurt, 37
Goethe, Johann Wolfgang von, 59, 71
Gold, Steven J., 73
Goldstein, Matt, 185
Gonzalez, E., 244
Grant, Madison, 30
Gray, Paul, 61
Great Depression, 12–3, 41, 53
Greico, Elizabeth, 97
Gropius, Walter, 37
Gutman, Herbert G., 30–1

Hadzide, Ruben, 119
Hall, Prescott F., 28
Halter, Marilyn, 73–4
Hansen, Marcus Lee, 97
Harvard University, 12
Hauptmann, Bruno Richard, 71
Hayek, Friedrich von, 37
Hebrew Immigrant Aid Society (HIAS), 47, 104
Heine, Heinrich, 59, 71
"here-there" connection, 254–5, 257, 264–5
Hernandez, Ruben, 179
Hershenson, Jay, 185
high school graduation rates, 167–74
Hindemith, Paul, 37
hip-hop, 121–2

Hirsch, Edith, 47
Hirsch, Julius, 47
Historically Black Colleges and Universities (HBCUs), 109, 123
Hitler, Adolf, 39, 54, 58
Holdaway, Jennifer, 89, 102
Hollinger, David, 107, 276
Holocaust, 45, 52, 54, 104
Holton, Gerald, 17, 35, 52, 57
Homeland Security Act, 238–40, 245–6
  see also United States Department of Homeland Security (DHS)
Hull House, 24
Hunter, Robert, 26

Ibanez, Blanca, 188n1
immigrant, definition of, 226
immigrants and refugees
  from Cambodia, 76, 82
  from Cape Verde Islands, 114–6, 124
  from China, 16–8, 82–3, 85, 90–2, 100n1, 105, 107, 151–2, 180–1, 212
  from Cuba, 76–7, 86, 211–3
  from the Dominican Republic, 90, 92, 98–9, 100n1, 105, 107, 118, 151–2, 179, 183, 255
  from Germany, 17, 28, 37–40, 43, 51–9, 63–5, 67–8, 71, 211, 227
  from Haiti, 123–4, 151–2, 211–3
  from Hong Kong, 82
  from Indonesia, 82
  from Italy, 7–21, 26, 97, 278
  from Korea, 16–8, 193
  from Laos, 76, 79, 82, 86
  from Liberia, 63, 67, 76, 86, 114–5, 122–3
  from Malaysia, 82
  from Mexico, 16, 72, 74, 99, 108–10, 129–47, 149, 151–2, 167–74, 175n2, 177–88, 188n2, 239–41, 256–63
  from Nigeria, 72, 114–5, 118, 120–2, 124
  from the Philippines, 82
  from Puerto Rico, 90, 92, 98–9, 100n1, 105, 108, 110, 179, 183
  from Russia, 9, 13, 15, 52, 57, 63, 67–9, 74, 76–8, 80–1, 86, 88, 90, 95, 101–6, 109
  from Senegal, 114, 120
  from Sierra Leone, 114–7, 122, 125
  from Somalia, 76–9, 86
  from the Soviet Union, 16, 52, 67–8, 73–81, 86, 100n1, 104, 192
  from Sudan, 73, 78, 86–7, 211, 214, 219
  from Thailand, 82
  from Vietnam, 73, 76, 78–86, 88, 117, 196, 211
  from West Africa, 67, 74, 113–25, 269
  from West India, 90–1, 99, 105, 107, 118, 120–1, 126
  see also Jewish immigrants and refugees; Mexican migrants
Immigration and Intergenerational Mobility in Metropolitan Los Angeles (IIMMLA), 133–4, 137, 140, 144, 146–7
immigration mind-set, 23
Immigration Reform and Control Act (IRCA), 130, 137, 139–40, 145–6
Immigration Restriction League, 28
in-between position of immigrants, 126, 261
income, 16, 39–40, 58, 62, 65, 78, 80, 83–4, 92, 99, 139, 143–6, 153, 168–71, 180–3
integration, 4, 29, 38, 41, 57–9, 107, 130–2, 145–6, 165, 197–205, 214–23, 225–34, 276
International Rescue Committee (IRC), 189, 193, 196–200, 206
Itzigsohn, Jose, 256, 265n1

Jeffersonian Research, 3
Jenks, Jeremiah W., 27, 30
Jewish Family Services, 165

Jewish immigrants and refugees, 9–20, 21n2, 31, 35, 39–40, 47, 63–4
   German, 17, 51–9, 64–5, 67–8
   Hungarian, 64
   Iranian, 52
   Russian, 9, 13, 15, 52, 63, 67, 74, 90, 95, 101–6, 109
   Soviet, 16, 76–88
Johnson, Violet, 73–4
Judaism, 52, 55–8

Kabba, Alex, 118
Kaiser Family Foundation National Survey of Latinos, 256, 258–63
Kamara, Sekou, 123
Kasinitz, Philip, 18, 73, 89
Kasson, John F., 24
Katznelson, Ira, 269
Kett, Joseph F., 26
Kim, Jane S., 189
Kim, Rose, 109
Kim, Susan, 18
Kindertransporte, 47

LaGuardia, Fiorello, 12
Lam, Tony, 84
language acquisition, 42–3, 48, 194, 234
   see also English as a Second Language/English Language Learners, ESL/ELL
Lara-Cinisomo, Sandra, 188n1
Lazarsfeld, Paul, 37
Leach, Mark A., 74, 130
legal permanent resident (LPR), 137, 139–40, 144, 244
legalization, 74, 130–4, 137, 139–40, 144, 145–7, 220
Levitt, Peggy, 255, 262
Longitudinal Immigrant Student Adaptation Study (LISA), 151, 157
Lutheran Immigration and Refugee Service (LIRS), 165, 189–90, 211, 227–8, 233
Lewin, Kurt, 57
Little Saigon (Orange County), 83
Lodge, Henry Cabot, 28

London, Jack, 25
Louima, Abner, 124

Madarocka, 121–2
Mandingo, *see* immigrants and refugees, from Liberia
Mann, Thomas, 37, 71
Marcuse, Herbert, 37
Martin, David, 76
Masjid Lwabahu (Gambian mosque), 117–8
Massey, Douglas, 108
Matthew effect, 5
McNees, M., 181
Merton, Robert K., 5
Mexican American Students Alliance (MASA), 184, 186
Mexican Consulate-Mexican Communities Program, Baruch College's, 177–9, 183, 186–7
Mexican Educational Foundation of New York (MexEd), 184, 186, 188n1
Mexican Mentorship Project, 183–5
Mexican migrants, 72, 130, 256–64
Midas, Eric, 68
migrant children, 130, 211–22, 225–35
migrant, definition of, 224n1, 225
Migration and Refugee Services of the United States Conference of Catholic Bishops (USCCB/MRS), 209–22, 224n2
Mill, John Stuart, 96
Miller, Herbert, 31
Miller, Mark J., 88
Min, Pyong Gap, 109
mobility, 4, 10–26, 31–2, 34n22, 78, 80, 94, 98–9, 116–9, 180, 194, 276–8
   downward, 26, 46, 90, 99, 269
   non-zero-sum, 269–70, 274
   upward, 12, 89–91, 102, 109–10, 116–7, 170, 174, 182–3, 251, 269–70, 278
Mogilescu, John, 188n4
Mollenkopf, John, 89, 102
Moynihan, Daniel Patrick, 97
multiculturalism, 93–7, 107, 254

National Socialism, European, 2, 35, 37–9
naturalization, 11, 74, 131–4, 139–40, 143–6, 263, 269
Neckerman, Kathryn, 109
Nee, Victor, 94, 254
New York Association for New Americans (NYANA), 104–5
New York City public schools, 81, 177, 179, 183, 185–7
New York Second Generation Study, 73, 89–99, 100n1, 104
Nigerian Women's Association of Georgia (NWAG), 118, 124
Nkrumah, Kwame, 115
No Child Left Behind (NCLB), 163, 194–5, 203–4
non-governmental organizations (NGOs), 42, 47, 67
nonprofit organizations, 124, 178, 179, 186–7, 189, 226–7, 229–30, 232–4, 248
non-zero-sum mobility, 269–70
see also mobility

Obama, Barack, 107
occupations, 13, 26, 39–48, 58, 78, 81, 83–4, 137–8, 143–5, 251, 268–77
Other than Mexicans (OTMs), 241

Panofsky, Erwin, 37
paradox of aid efficiency, 48
Park, Robert, 31, 178
Patterson, Orlando, 14
Paul and Daisy Soros Foundation, 103
Peck, Abraham J., 58–9
Perlmann, Joel, 11, 13, 15, 149, 255, 266n1
Pew Hispanic Center National Survey of Latinos, 256, 258–63
Pew Hispanic Center National Survey of Mexican Migrants, 256–763
Pilot National Asian American Political Survey (PNAAPS), 256–63
Piscator, Erwin, 37
political participation, 257, 260, 262–3

Portes, Alejandro, 91, 101, 255
poverty, 25–6, 62, 76, 98, 105–6, 109, 111, 125n3, 153–7, 161–2, 173, 195, 197
Progressive era, 29–32
Project Second Wave, 2, 51
psyche for success, 74, 113, 115–6, 122

Quakers, 47
quotas, 28–9, 53, 269

racial stereotyping, 19
racism, 93, 108–10, 121, 195, 269, 276
Refugee Act of 1980, 3, 76, 192
refugee camps, 76, 81–3, 86–7, 196–7, 205, 216–20, 223, 227, 231
refugee, definition of, 206n4, 226
Reucker, Anne, 188
Rhodes, Jean, 184
Richards, Orabella, 123
Riis, Jacob, 25, 27, 30
Ritterband, Paul, 80
Roberts, Joy, 122
Rodriguez, Richard, 110
Rosenzweig, Franz, 55
Rumbaut, Ruben, 91, 101
Russell Sage Foundation, 3, 5, 25, 102, 133, 147

Saigon, 82, 192, 211
Salamon, Lester M., 179
Sarukhan, Arturo, 185
Saucedo, Silvia Giorguli, 256, 265n1
Schiller, Friedrich, 59, 71
Scholem, Gershom, 55–6
Schurman, Jacob Gould, 28–9
Second Generation in Metropolitan New York, 102
Second Wave refugees, definition of, 51–2
Shakespeare, William, 71
Shaler, Nathaniel, 28
Simmel, Georg, 96
Siulc, N., 292
Smith, Robert, 109, 149, 181
social mobility, *see* mobility

Society of Friends (Quakers), 47
Sonnert, Gerhard, 17, 35, 52, 57
Soros Foundation, 103
Special Immigrant Juvenile Status (SIJS), 244, 249
Spencer, Herbert, 27
Strong, Josiah, 28, 30
Strauss, Leo, 57
student-immigrant tradition, 116
students with interrupted formal education (SIFE), 195, 199–200, 203–5, 208n33
Suárez-Orozco, Carola, 149
success
  causes of, 41–8
  definition of, 4
  drive toward, 45
  framework for, 228–9
  inhibitions to, 218–21
  psyche for, 115–6
  socioeconomic indicators of, 39–41
  sociology of, 9
Sudanese Lost Boys, 73, 86–7, 211
Sumner, William Graham, 27
supplemental security income (SSI), 105–6
Szilard, Leo, 37

Taylor, Charles, 67
Thomas, W. I., 31
Tocqueville, Alexis de, 3
Tokely, James, 123
Tota, Kenneth, 190
trafficking, human, 191, 213–8, 222, 230, 240–48
transnationalism, 27–8, 124, 132, 251, 253–6, 261–4, 265–6n1
Tsao, Hui-shien, 73, 251

Ueda, Reed, 7, 19
Ukrainian refugees, 76, 86
Umoja Media Project, 124
unaccompanied alien children (UAC), 237–49
United Nations High Commissioner for Refugees (UNHCR), 76, 219

United States Citizenship and Immigration Services (USCIS), 244
United States Conference of Catholic Bishops (USCCB), 189, 209–22, 224n2, 228
United States Department of Homeland Security (DHS), 240–1, 244–6
  see also Homeland Security Act
United States Department of Justice (DOJ), 238, 240
United States Immigration and Customs Enforcement (ICE), 68
United States Immigration and Naturalization Service (INS), 238, 240–1, 246
United States Office of Refugee Resettlement (ORR), 3, 87, 192, 203, 228, 237–48
United States Refugee Admissions Program (USRP), 192

Vermeulen, Hans, 14
Vickerman, Milton, 118
Vietnam-China conflict of 1978, 82

Waldinger, Roger, 251
Walker, Francis A., 26
Washington, Booker T., 172
Waters, Mary C., 73, 102, 118, 255
Weill, Kurt, 37
Weisskopf, Victor, 37
Wertheimer, Jack, 54–5
Wilder, Billy, 37
Wilson, Annie, 189
Wilson, William Julius, 108

xenophobia, 31, 70

Youth for Sierra Leone, 125

Zhou, Min, 17–8
Zionism, 56
Zuniga, Victor, 179

GPSR Compliance
The European Union's (EU) General Product Safety Regulation (GPSR) is a set of rules that requires consumer products to be safe and our obligations to ensure this.

If you have any concerns about our products, you can contact us on

ProductSafety@springernature.com

In case Publisher is established outside the EU, the EU authorized representative is:

Springer Nature Customer Service Center GmbH
Europaplatz 3
69115 Heidelberg, Germany

www.ingramcontent.com/pod-product-compliance
Lightning Source LLC
LaVergne TN
LVHW011800060526
838200LV00053B/3636